PROGRAMMABLE CONTROLLERS

PROGRAMMABLE CONTROLLERS
A Practical Guide

Denis Collins
Eamonn Lane

McGraw-Hill Book Company

London · New York · St Louis · San Francisco · Auckland · Bogotá · Caracas · Lisbon
Madrid · Mexico · Milan · Montreal · New Delhi · Panama · Paris · San Juan · São Paulo
Singapore · Sydney · Tokyo · Toronto

Published by
McGRAW-HILL Book Company Europe
Shoppenhangers Road, Maidenhead, Berkshire SL6 2QL, England
Telephone 01628 23432
Fax 01628 770224

British Library Cataloguing in Publication Data
Collins, Denis A.
Programmable Controllers
I. Title II. Lane, Eamonn J.
629.895

ISBN 0-07-709017-9

Library of Congress Cataloging-in-Publication Data
Collins, Denis
Programmable controllers / Denis Collins, Eamonn Lane.
p. cm. 98
Includes bibliographical references and index.
ISBN 0-07-709017-9 (alk. paper)
1. Programmable controllers. I. Lane, Eamonn
II. Title.
TJ223.P76C65 1995
629.8′95 – – dc20

First published in Ireland by Collins and Lane, Cork, Ireland.

Reprinted 1997

Typeset by Focal Image Ltd, London
and printed and bound in Great Britain at the University Press, Cambridge

Printed on permanent paper in compliance with ISO Standard 9706.

To Loretto and Noreen

Contents

Acknowledgements

At the conclusion of over a year's research and writing effort we are keenly aware that many individuals and groups have contributed to the making of this book.

To our colleagues at the Cork Regional Technical College we offer our thanks for their encouragement and support. We are especially indebted to Richard Daly for his many helpful suggestions and comments.

Since we set out to make the book as relevant as possible to learners and users, it was essential that it contain references to real products. We wish to thank the following companies for permission to use their company names and product information and for providing every conceivable support and assistance.

- Mitsubishi Electric
- Siemens
- Sprecher+Schuh
- Telemecanique
- Brodersen Control Systems

We are especially grateful to many individuals for their patience, suggestions and comments, always delivered constructively and professionally. Our thanks to

- Fergus Madigan, Cíaran Moody, Tom Sheridan (Mitsubishi Electric)
- Liam Mulligan, Roger Muller, John McPhillips (Siemens)
- Peter Cox, Joe McCarthy, Tom Weafer, Pat Horgan (Sprecher+Schuh)
- Jean-Pierre Mura, Pat Carr, Jim Rice, James Fergus, Mark Keogh, Dómhnall Carroll (Telemecanique)
- Kurt Serup (Brodersen Control Systems)
- Louise Gibney (Philips Electronics)
- Ray Weldon (Galway RTC)
- Frank Turpin (Intel)

We also thank the editorial team of McGraw-Hill for their patience and guidance.

Trade marks

Preface

The programmable controller was introduced in the early 1970s in the US car industry and was an almost instant success. Early fears about its reliability and safety quickly gave way to widespread acceptance. Its obvious advantages assured its ready marketing in nearly every industrialized country. There are millions of units in service throughout the world today which testify to its functionality, reliability, flexibility and cost-effectiveness.

Development over the past decade has been hectic. Each new model has brought greater power and ease of use. More and more functions are squeezed into smaller and smaller packages, so that the automation of small and medium-sized projects is now truly affordable.

Two key elements explain the programmable controller's success: (a) it is accessible to a broad range of engineering personnel, and (b) it implements its control functions by means of a program. Thus it is possible for non-specialists to assemble sophisticated and complex control systems, the functions of which are limited only by human intellect and not by traditional physical barriers.

The main obstacle to the fullest possible benefit of programmable control is one of education and training. The sheer pace of change has opened a gap of up to 10 years between the technology developers and the broad mass of end users, some of whom feel alienated or even intimidated. This hinders the full acceptance of programmable control and discourages new users.

We hope our book will address this problem. We have been particularly careful to avoid a rigorous approach and we make no reliance on the reader's knowledge of computers. The numerous worked examples show how control problems can be tackled on four different programmable controllers, revealing the common principles and functions involved. The five graded, fully tested and documented projects give the reader a broad experience of the techniques of implementation and reflected our conviction that learning by example is highly successful. The 'soft' treatment of some of the more advanced functions will, we hope, help to broaden the horizons of existing users.

D.C.
E.L.

Graphical symbols

Electrical and electronic graphical symbols

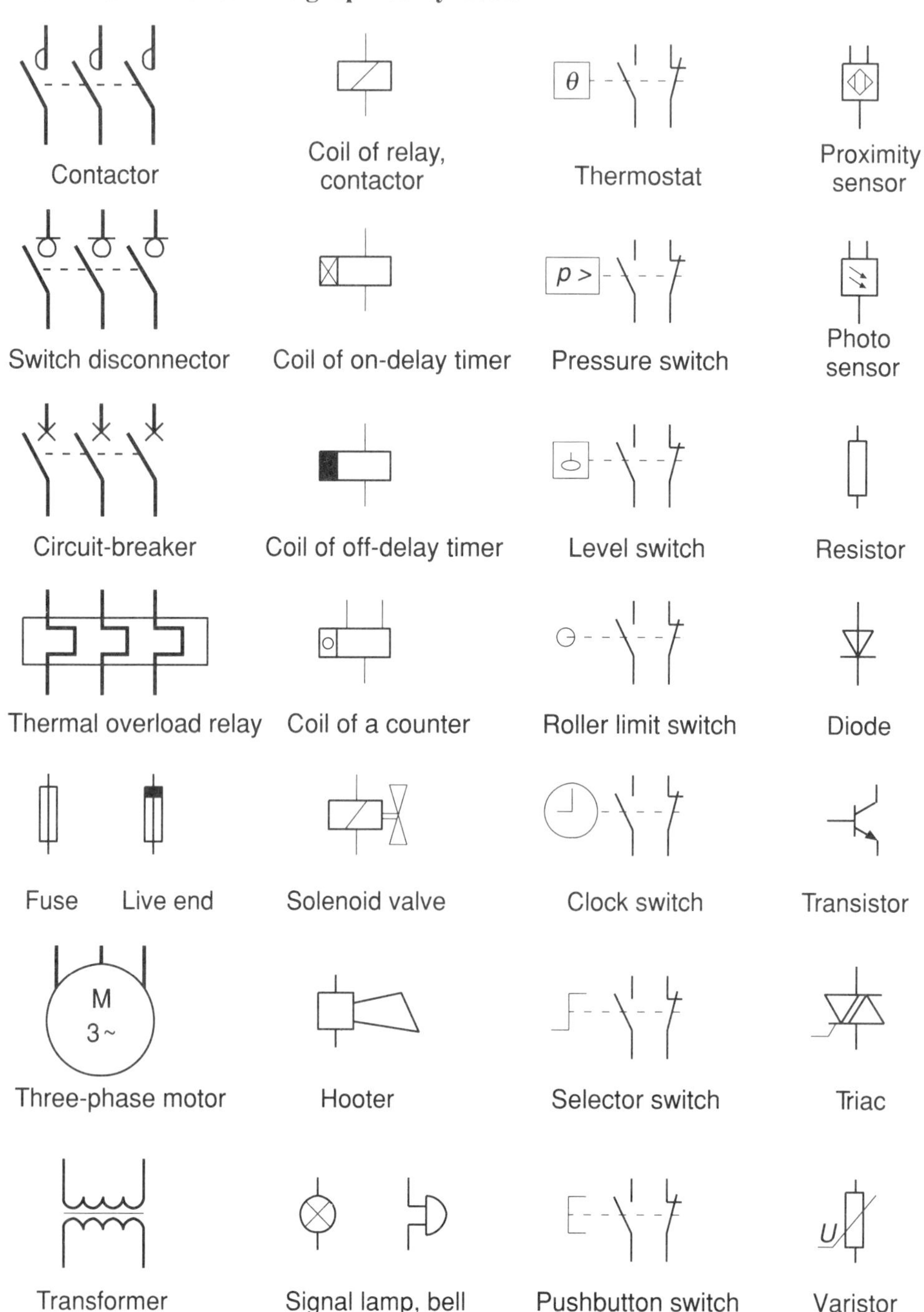

Note: With the exception of the photo-sensor, all symbols are in accordance with IEC 617.

Logic symbols

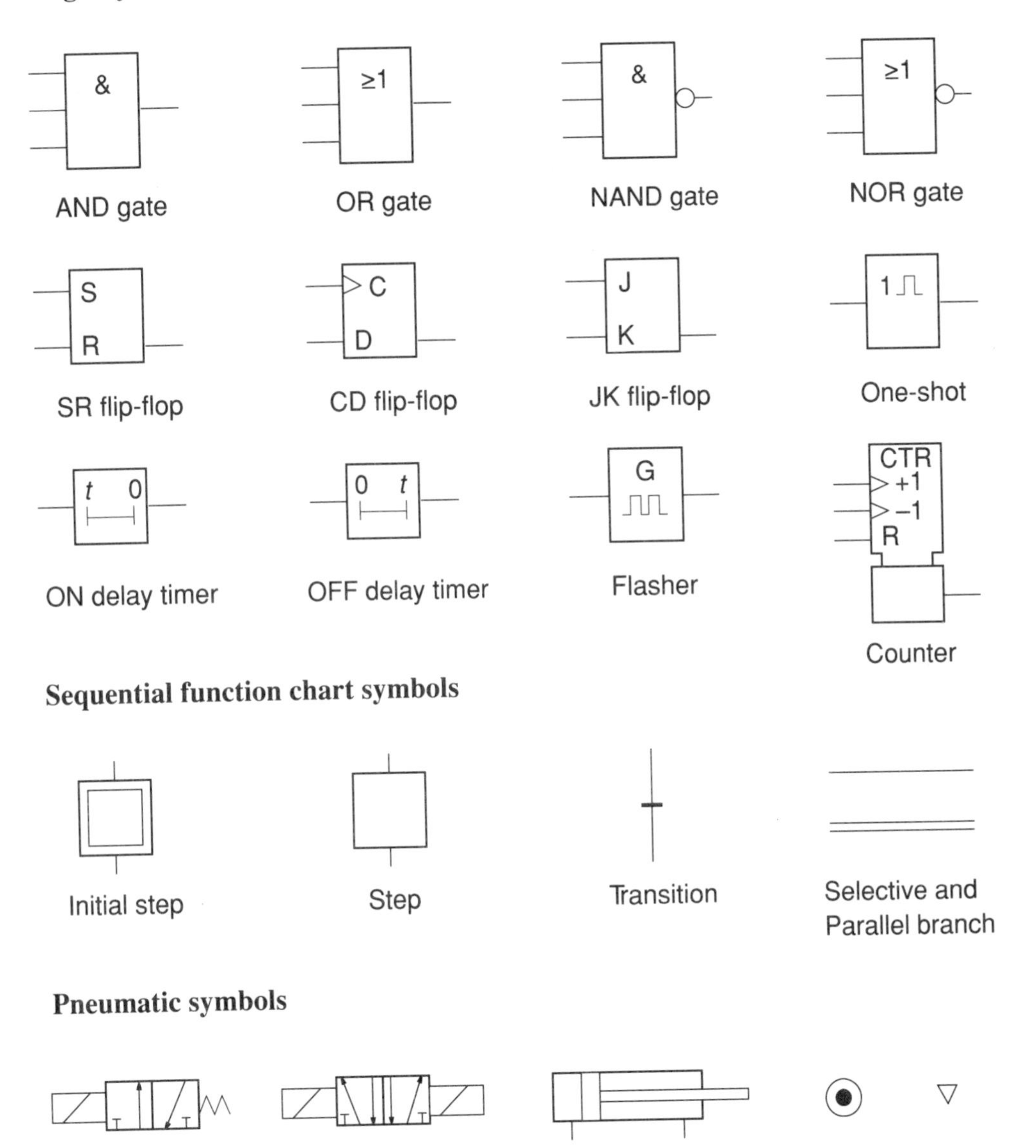

Note: Logic symbols are in accordance with IEC 617.
Sequential functional chart symbols are in accordance with IEC 848 and IEC 1131-3.
Pneumatic symbols are in accordance with ISO 1219.

Chapter 1 Introduction

1.1 Ten fundamental questions

1. What is a programmable controller?
A programmable controller is a microprocessor-based industrial controller, the functions of which are determined by a stored program.

2. What is a program?
A program is a set of instructions 'telling' the controller how to behave. It is stored in the controller's memory.

3. How does a programmable controller differ from a computer?
A computer is optimized for calculation and display tasks and is programmed by specialists. A programmable controller is optimized for control and regulation tasks and can be programmed by non-specialists. It is also well adapted to the industrial environment.

4. Why are programmable controllers so common?
Because they are cost-effective and have significant advantages over traditional control systems based on relays or pneumatics.

5. Where are they used?
In virtually every industry where automation is involved, from individual machines to whole processes, in commercial, institutional and industrial premises. The newest 'microcontrollers' are so cheap and compact that they are economical even in domestic applications.

6. What are the main advantages?
A control system based on a programmable controller is flexible, reliable and compact and can be assembled at a relatively low cost.

7. Are all programmable controllers the same?
They are broadly similar in a functional sense, but they differ in size, programming detail and mechanical design. Most manufacturers offer several models with different levels of performance to cater for the wide variety of tasks to which programmable controllers can be applied.

8. What tasks does a programmable controller perform?
The control tasks previously undertaken with electrical and/or pneumatic controls, e.g. interlocking, sequencing, timing and counting. It can, in addition, perform a variety of calculation, communication and monitoring tasks far outside the competence of traditional systems.

9. Does a programmable controller eliminate contactors and valves?
No, but these items are brought under the programmable controller's influence in modern control systems.

10. Are there drawbacks?
Yes. Programmable controllers still do not enjoy the same trust or acceptance as traditional control techniques, even though the technology is nearly 30 years old. The natural reluctance to accept the 'new' technology is understandable; most of our current industrial staff were educated and trained before this technology became common. Some technical adaptations have to be made in implementing programmed control. These problems will be overcome with

- education for new users and the subsequent building of experience
- a measure of understanding on the part of vendors and the 'converted'.

1.2 Control system overview

There are three characteristic features of a control system, whether it is programmable or not:

1. there are certain *actions* to be taken (such as turning a valve ON or OFF or regulating its position)
2. there are certain *rules* governing those actions—the control system should, as far as possible, be predictable and not subject to random behaviour
3. the rules take account of certain relevant conditions in the plant (such as manual switches, sensors for level, pressure, temperature, position).

It is interesting and informative to compare a control system of traditional design with one of modern design (using a programmable controller), in the context of these three features.

Figure 1.1 shows the two control systems applied to the same task, namely control of a relay K1. The *action*, in this case, is turning on K1. There are three switches in the plant, whose status is relevant to the control of K1. For simplicity we shall say that these three switches account for all the *conditions* that have to be checked. By simply examining the illustration we can see that the 'action' content and the 'condition' content of the two control systems are identical.

We must therefore conclude that the main difference between the two systems is in the implementation of the *rules*. In the traditional control system the wiring inside the control cabinet connects the three switches in series. This means that S1 and S2 and S3 must close for K1 to be turned on—the *wiring* makes the rules!

In the modern control system the place of the wiring is taken by a program. The instructions contained in the program must have the following effect: 'when S1 is closed AND S2 is closed AND S3 is closed, turn on K1'. To manage this, the programmable controller must be able to check the state of the three switches; this is facilitated by connecting each switch

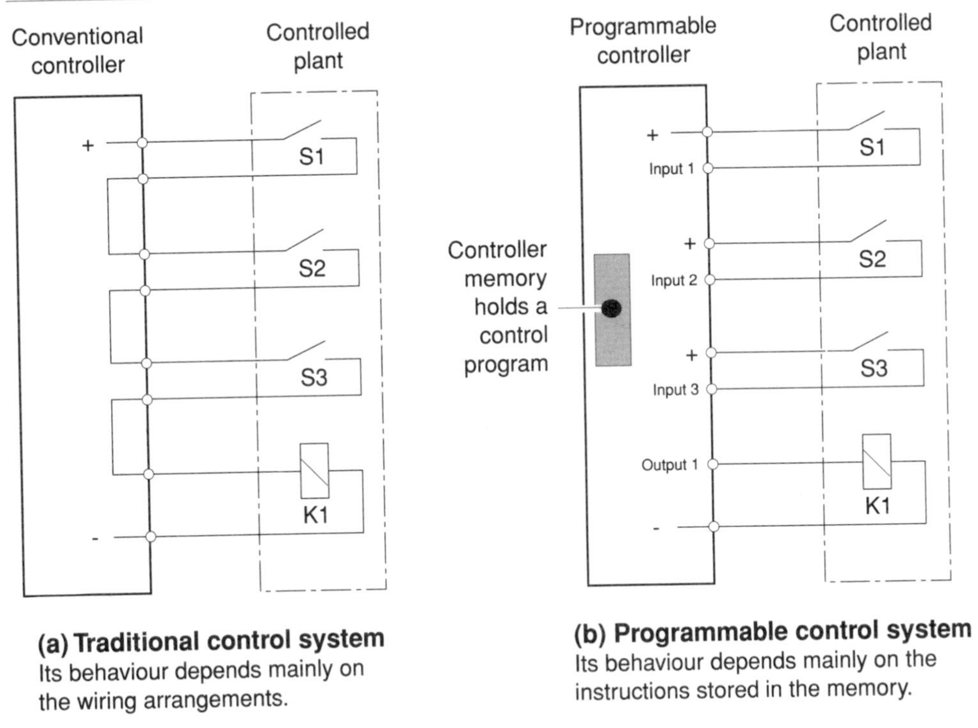

Figure 1.1. Comparing traditional and programmable control systems.

to an input terminal on the controller. It must also be able to take charge of the relay; this is facilitated by connecting the relay to an output terminal on the controller.

1.3 The benefits of programmable control

At first sight the solitary difference between the two systems may not seem profound but consider the following results:

Change of rules — If the rules need to be changed, a traditional system must be rewired. This may be inconvenient, expensive and protracted.
A programmable controller can be *reprogrammed* to accommodate a change of rules—no rewiring is needed. Neither is there any need to alter drawings, since program development systems have automatic documentation!

Extra functions — If extra functions are needed, a traditional control system must be fitted with the additional equipment—assuming the space is available for it.
A programmable controller has a vast array of built-in functions such as relays, timers, counters and sequencers which are freely accessible at any time and demand no extra space in the control cabinet.

Reliability

Moving parts are subject to mechanical failure, and mechanical failure accounts for a substantial proportion of faults in the components of traditional control systems. Programmable controllers have very few, if any, moving parts. In addition they are amenable to manufacturing methods that include exhaustive, automatic test procedures. Defective or even potentially defective components can be identified and excluded. As a result, reliability is excellent.

Communication

Traditional control systems have little or no potential for interconnection with other systems for the purposes of control, supervision or reporting. Programmable controllers are inherently suited to, and increasingly better prepared for, such a role. The availability of communications modules allows the connection of controllers to industrial networks, which facilitates data interchange on a grand scale.

1.4 Structure

Figure 1.2 illustrates the structure of the programmable controller and its setting in the control environment. The key piece of equipment is the microprocessor, which is the 'decision-maker' in the scheme. It makes decisions based on the instructions stored in the

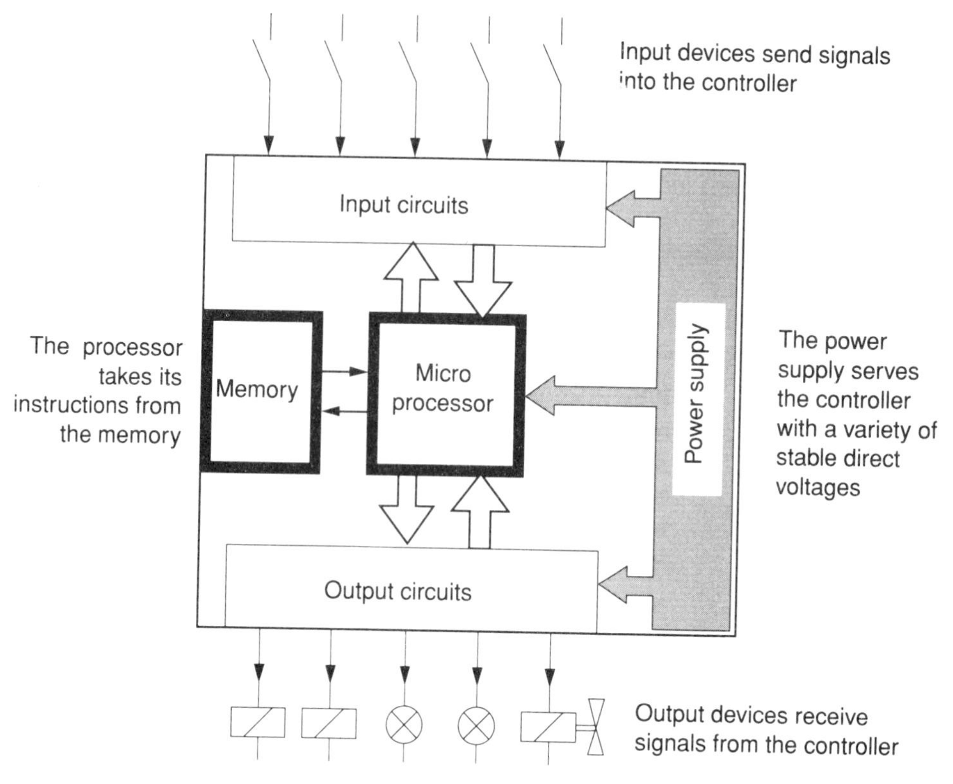

Figure 1.2. Fundamental structure.

memory (the program). Nowadays, microprocessor chips are very small, reliable, powerful and, being mass-produced, they are relatively cheap.

The switches, detectors and sensors in the plant are connected to the input terminals of the controller, where the microprocessor can evaluate their status. These input devices send signals *into* the controller. The only task of the input circuits is to interface between the input devices and the processor. This means isolating them electrically and adjusting for the voltage difference between them. Thus the processor can read the input signals without having to tolerate the high plant voltages.

The relays, valves and indicators in the plant are connected to the output terminals where the microprocessor can exercise its control influence. These output devices receive their signals *from* the controller. The output circuits are there to provide isolation and to adjust for the voltage difference between the processor and the output devices. Thus the processor can transmit the output signals without having to encounter the (relatively) high plant voltages. The power supply serves the various parts of the controller with the right kind and rating of supply. It may also supply the voltage for the input circuits, but not the output circuits.

1.5 The processor

The processor fitted in a modern programmable controller is a class of semiconductor chip known as an embedded microcontroller. All the functions of a computer are encapsulated on this chip, which typically has an active area of less than 1 cm^2 and contains hundreds of thousands of transistors (Fig. 1.3). With the support of other chips, it is capable of executing many millions of instructions every second. It communicates with the memory, input and output interfaces through a system of conductors called a *bus*, on which it may send/receive 8, 16 or 32 signals at a time. The operating voltage is typically 5 V d.c.

Modern semiconductor products are subject to a rigorous design process, a manufacturing technique of extraordinary cleanliness and accuracy, and the most intensive testing. As a result, they are extremely reliable in service and have a failure rate less than 10 per cent of the rate for the next best technology.

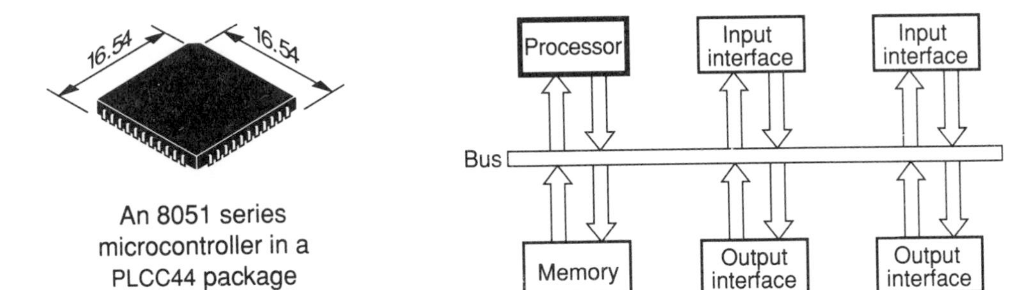

Figure 1.3. The microprocessor.

1.6 Exercises for Chapter 1 (Solutions in Appendix 1)

1.1. Why is the microprocessor sometimes called the brain of the programmable controller?

1.2. What tells the programmable controller how to behave?

1.3. To what terminals of the programmable controller are the plant control switches connected?

1.4. Why are programmable controllers so reliable?

Chapter 2 The typical programmable controller

2.1 The main features

We shall begin by looking at a 'typical' controller and describing its salient features (refer to Fig. 2.1). Every controller has three sets of terminals through which it connects to the plant. These are

- power terminals
- input terminals
- output terminals.

The power terminals simply refer to the connection of the mains supply (the controller needs power to function). In most modern programmable controllers provision is made for 115 V a.c. or 230 V a.c. supplies, and some are even self-adjusting within that range. Even though the current demand of the controller itself is very modest (typically a few mA), a substantial protective conductor is always used.

The input terminals are used for the connection of the switches and sensors in the plant. These so-called input devices send signals into the controller, allowing it to 'see' the status of various parts of the plant. Examples (Fig. 2.2) of input devices are

- contacts of manual control switches (pushbuttons, selectors, etc.)
- contacts of automatic control switches (for level, pressure, temperature, etc.)
- contacts of relays or contactors
- position sensors (proximity, photoelectric, etc.).

Each input device is individually connected to one input terminal on the programmable controller.

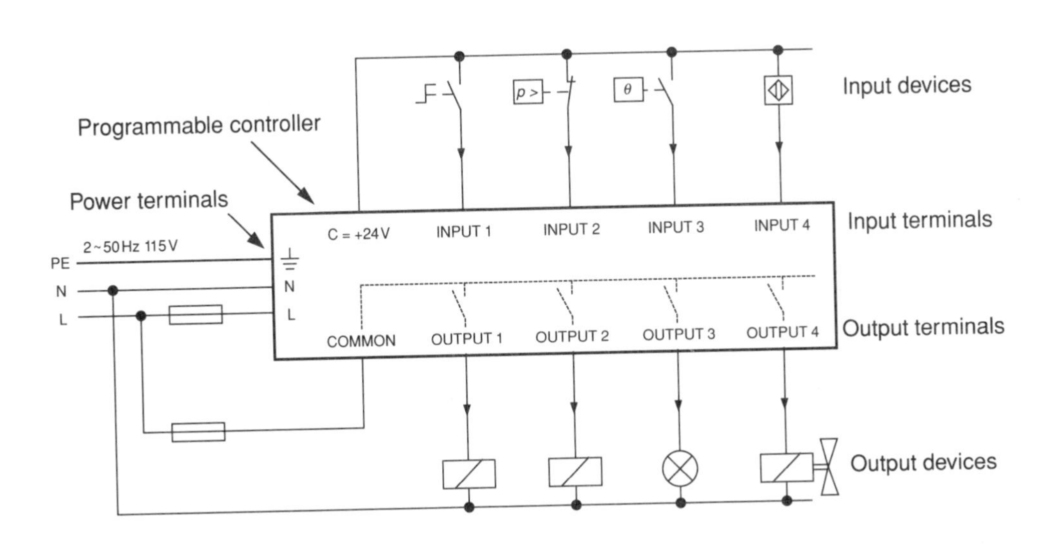

Figure 2.1. Electrical connections to a programmable controller.

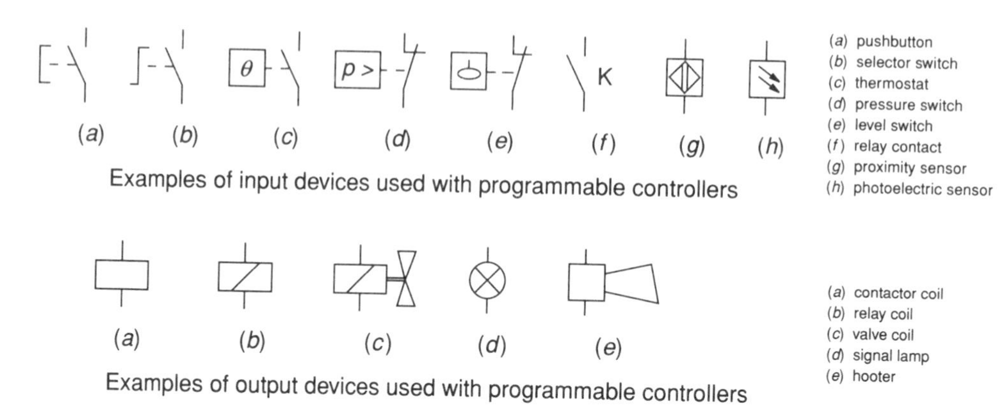

Figure 2.2. Typical input and output devices.

The output terminals are used for the connection of any item in the plant that has to be controlled. The programmable controller exercises control over such items by turning their supply ON or OFF or by regulating it. Examples (Fig. 2.2) of output devices are

- coils of contactors or relays
- coils of solenoid valves (pneumatic or hydraulic)
- indicators (signal lamps or sounders).

Each output device is individually connected to one output terminal on the programmable controller. Figure 2.1 shows a selection of input and output devices connected to a programmable controller . All the inputs operate at 24 V d.c.—a common but by no means exclusive arrangement. The output devices operate at 115 V a.c. and receive supply through sub-miniature relays inside the programmable controller. When a particular output, say OUTPUT 3, is turned on, the corresponding relay closes its contact and connects the supply to the signal lamp.

2.2 Mechanical design

A variety of mechanical designs has evolved over the years, the most common of which are the *block* type and the *rack* type, both illustrated in Fig. 2.3. The block type is common to many 'small' controllers and comes as an integrated package complete with power supply, processor and a fixed number of input and output terminals in a compact box. It has advantages where shallow mounting depth or easy access to terminals is important.

The rack type is used for all sizes of controller. The various functional units—power supply, processor, inputs and outputs—are packaged into individual *modules* which can be plugged into sockets in the base unit. The user decides the mix of modules best suited to his or her requirements and assembles them accordingly. It has advantages where large numbers of inputs and outputs are involved, where a variety of input and/or output voltages has to be catered for, where the ratio of inputs to outputs is unusual or where rapid replacement of faulty cards is needed.

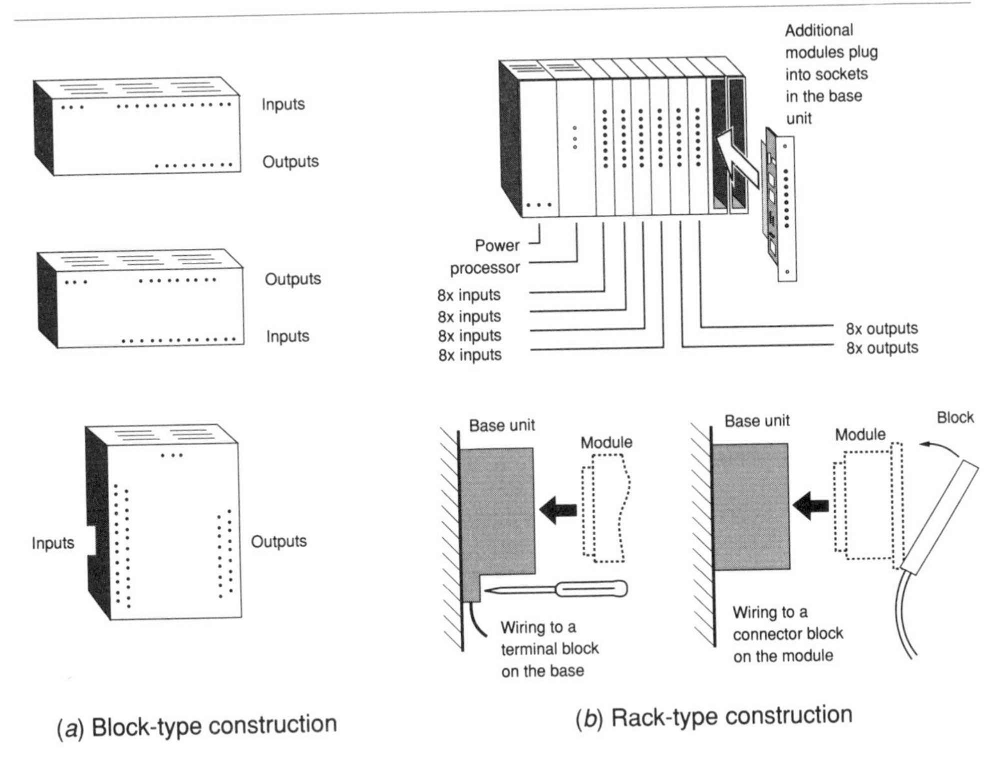

Figure 2.3. Common mechanical designs.

2.3 I/O addresses

Each input terminal and output terminal has a unique name or number, which is designated by the manufacturer of the controller. This is its so-called I/O 'address', and is the means used by the controller to identify the signal of an input or output device. The user will notice differences of addressing between the various manufacturers, since there is little standardization in this area as yet. For now, the user must adapt to the differences. This book is based on four controllers of different manufacture; their addressing systems are broadly representative and easily learned.

The four controllers (in alphabetical order) are

- Mitsubishi F_2-40MR-ES and Fx-48MR-ES
- Siemens SIMATIC® S5-100U
- Sprecher+Schuh SESTEP® 290
- Telemecanique TSX 17_20.

Their I/O arrangements are illustrated in Fig. 2.4. All four controllers are profiled in greater detail later in this chapter.

To demonstrate the addressing differences, let us take a simple example. Suppose we have an installation with three switches S1, S2 and S3 and one signal lamp H1 (Fig. 2.5(*a*)), and we want to connect them to the first available input and output terminals on the

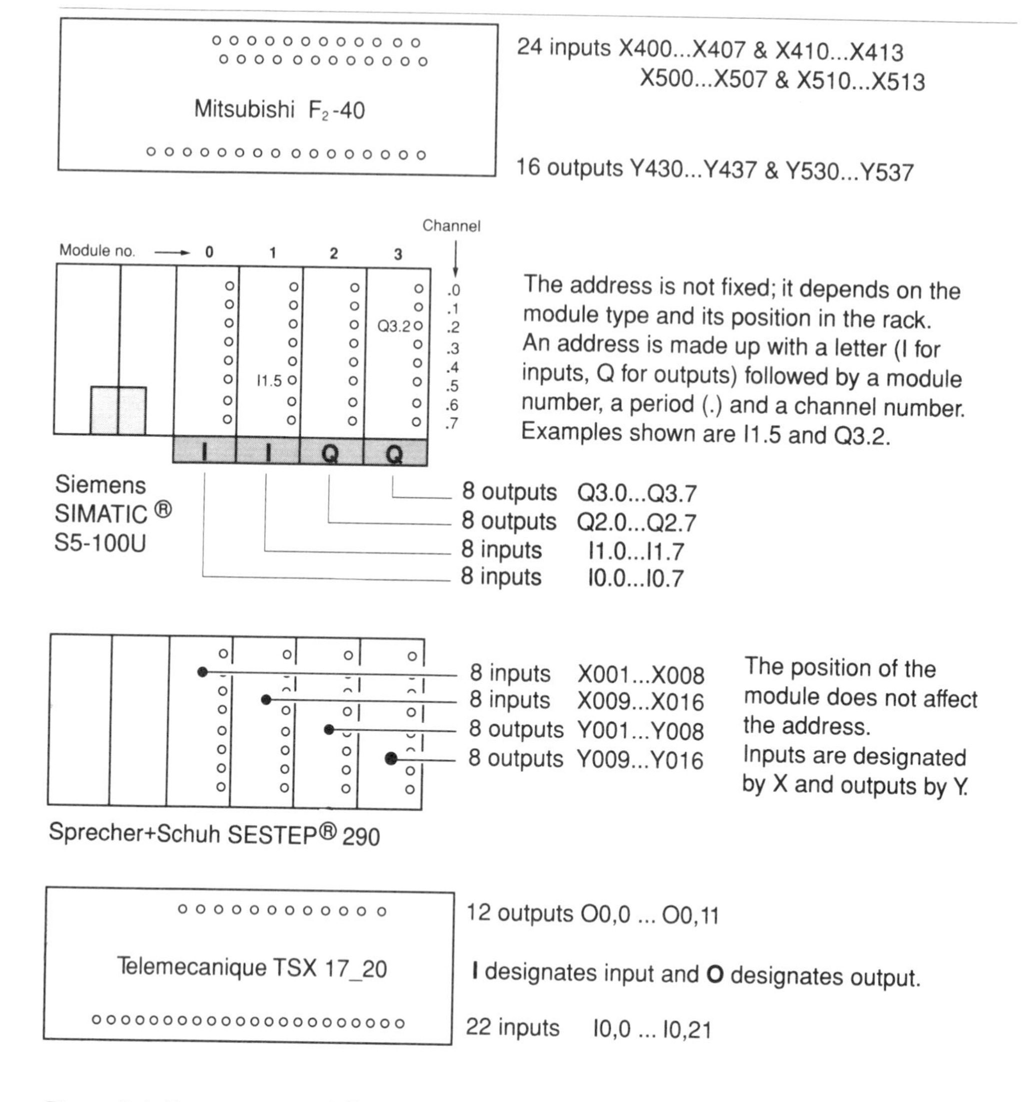

Figure 2.4. Front view and I/O addresses on four programmable controllers.

programmable controller. There are four solutions, one for each controller, but the only differences are the I/O addresses as shown in Fig. 2.5(*b*)–(*e*).

In the above example we specified the 'first available terminals'; note however that in a real project the user is free to choose which input or output terminal is to be used for any particular item.

The procedure of declaring the particular inputs and outputs to be used and the items to be connected to them is an important task which must be carried out early in every exercise or project. It is known formally as the *I/O assignment*.

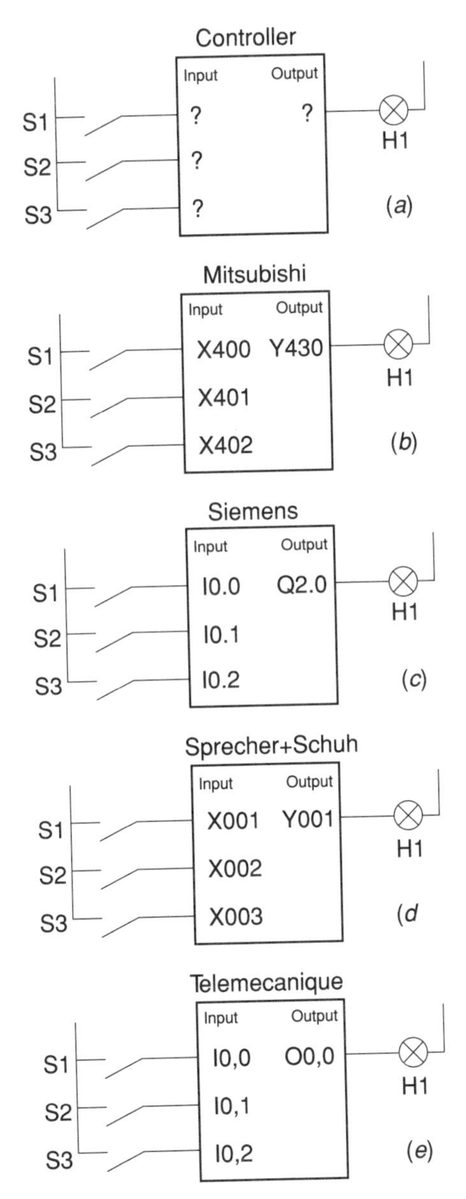

Figure 2.5. Addressing differences.

2.4 The program

As mentioned above, the program is a set of instructions that determines the control behaviour of the programmable controller. It is created by (or for) the user with the help of a programming terminal. Its main purpose is to set down the precise conditions for turning on (or regulating) each output of the controller.

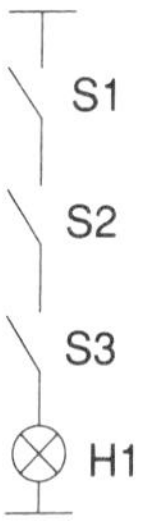

S1

S2

S3

H1

(*a*)

(*b*) Mitsubishi		(*c*) Siemens		(*d*) Sprecher+Schuh		(*e*) Telemecanique	
LD	X400	A	I0.0	STR	X001	L	I0,0
AND	X401	A	I0.1	AND	X002	A	I0,1
AND	X402	A	I0.2	AND	X003	A	I0,2
OUT	Y430	=	Q2.0	OUT	Y001	=	O0,0

(*a*) Circuit diagram, the design criterion in this example.
(*b*)–(*e*) Programs for making the controllers perform like the circuit diagram.

Figure 2.6. Circuit diagram and the corresponding control program for four controllers.

As an example of a simple piece of programming, consider the circuit diagram in Fig. 2.6(*a*). It comprises three switches connected in series to the signal lamp. Clearly, all three switch contacts need to be closed for the lamp to function.

To get a programmable controller to do the same job as this circuit, we need to wire the switches to inputs and the lamp to an output on the controller. This is already done in Fig. 2.5. We now need to equip the controller with a program that will instruct it to 'connect' the three inputs in series to turn on the output for the lamp. The necessary program is given for each controller in Fig. 2.6 in the so-called list form.

It is noteworthy that the instructions used for programming the different controllers are not identical, although there are obvious similarities. The effect of all four programs is roughly as follows:

- if there is a signal at the first input
- AND a signal at the second input
- AND a signal at the third input
- then turn on the first output.

Programming techniques are treated in greater detail in Chapter 4.

2.5 Running the program

The programmable controller has two main operating modes, STOP and RUN. In stop mode the controller is powered up but is not performing any control functions (rather like a car with its engine idling). In the run mode it executes all the instructions contained in the memory. The mode can be changed by using

- externally connected STOP/RUN switches (Mitsubishi)
- integral STOP/RUN toggle or key switch (Siemens)
- STOP/RUN commands on the programming terminal (Sprecher+Schuh and Telemecanique).

The program is stored in the memory in the form of statements, a few of which we encountered in the last example. Each statement in the memory is given its own space, called a step. Every step is numbered so that the instructions can be treated in a definite order and recalled when necessary. Figure 2.7 shows a typical memory configuration (Mitsubishi instructions are used but the principle is universal).

Memory content

STEP No	INSTRUCTION	ADDRESS
0000	LD	X400
0001	AND	X401
0002	AND	X402
0003	OUT	Y430
0004	LD	X403
0005	AND	X400
0006	OR	X405
0007	OUT	Y431
0008	LD	X400
0009	OR	X410
[illegible]	OR	X40[illegible]
1998		
1999		

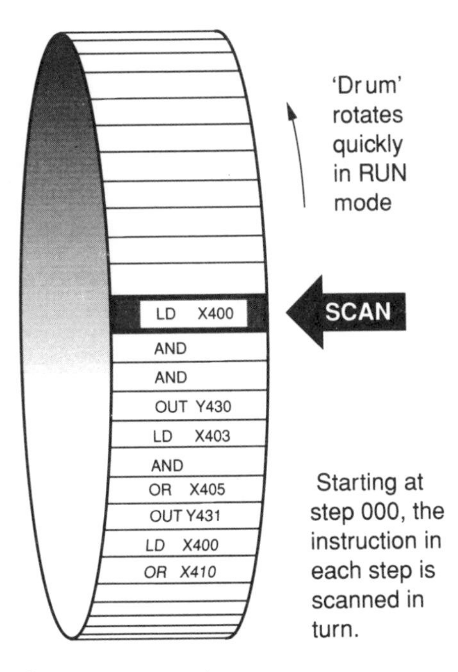

Figure 2.7. Visualizing the program running.

To help visualize what happens in run mode, the instructions are imagined to be mapped on to a drum rotating at high speed. The 'stationary' processor scans each instruction in quick succession and logically assembles a list of outputs to be turned on. (The processor's weakness is that it deals with one instruction at a time; its strength is that it does so *extremely* quickly.)

This whole procedure of scanning inputs, outputs and the program and finally updating the outputs is known as a *cycle*. While it is in the run mode, the controller must repeat this cycle indefinitely and very quickly if it is to remain in control of the plant. The problem with a slow cycle is that a brief input signal appearing at an unfavourable moment could be completely missed by the processor and lead to a malfunction. For most modern controllers a typical cycle time could be less than one hundredth of a second.

Every controller incorporates a protective mechanism (known as a *watchdog timer*) which operates automatically if the cycle time is excessive. If this device operates, the program is stopped, all outputs are turned off and a warning lamp on the front of the controller is illuminated.

2.6 Limits

Every controller is limited in two fundamental ways: memory size and the maximum number of I/Os.

Memory size is generally described as a certain number of *bytes*. A byte typically holds one instruction (an instruction such as 'STR X004' in the Sprecher+Schuh controller). Clearly the memory must be adequate to store every instruction needed for a working program and ideally have something to spare to provide for changes or extensions. For example, the standard memory supplied with the Siemens SIMATIC® S5-100U controller (CPU 100) is 2048 bytes (or 2k, where k = 1024 bytes).

A declaration by a manufacturer as to the 'max I/O' of a controller indicates the largest number of combined inputs and outputs that the controller can manage directly. For example, the stated figure for the Telemecanique TSX 17_20 is 160.

The following understanding of the meaning of 'small', 'medium' and 'large' has thus emerged without any formal definition

Small	Up to 128 I/O, 2k memory and destined to deal with control tasks of moderate complexity (machine control, relay replacement).
Medium	Up to 512 I/O, 16k memory and destined to deal with control tasks of significant complexity (machine and process control, analog functions, etc.).
Large	Up to 4096 I/O, 96k memory and destined to deal with major control tasks (entire processes, including calculation, supervision and monitoring operations).

Details of some of the controllers proposed in the exercises and case studies in this book are given in Figs 2.8, 2.9, 2.10 and 2.11, together with short profiles of the manufacturing companies. See pages 16–19.

2.7 Exercises for Chapter 2 (Solutions in Appendix 1)

2.1. What types of switches or detectors are suitable for detecting the position of a moving part of a machine?

2.2. What are the usual control voltages used in industrial plants?

2.3. With a rack type of programmable controller, what modules are normally required for any simple control task?

2.4. What is the minimum number of I/Os on the programmable controller for a plant with 12 control switches and 8 motor contactors?

Details of the controllers featured in this book, together with short profiles of the manufacturing companies now follow.

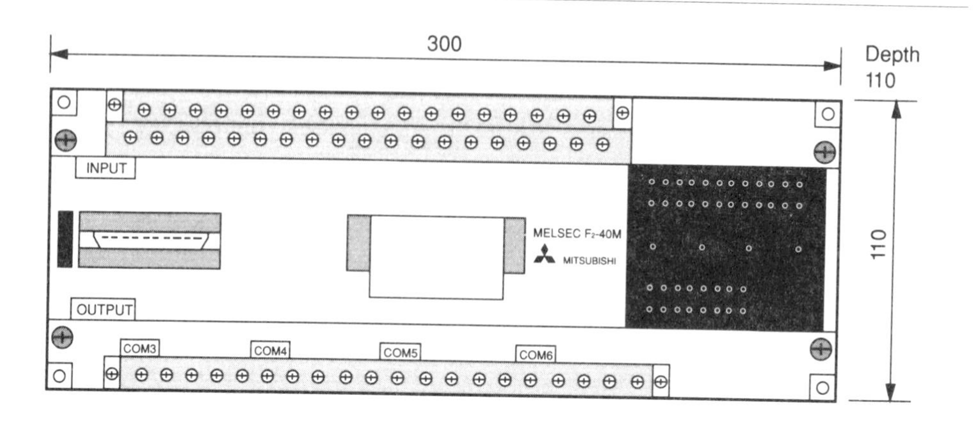

Figure 2.8. The Mitsubishi F_2-40MR programmable controller (dimensions in mm).

2.8 Company and product information

Company

Name	Mitsubishi Electric Corporation
Headquarters	Mitsubishi Denki Building, Marunouchi, Tokyo 100, Japan
Main business activities	Industrial, electronics and consumer products
Employment world-wide	112 000 (1994)

Product

Model	F_1/F_2/Fx series programmable controller	
Manufactured	Himeji Works, Japan	
Year introduced	1985/1993	
Production	>360 000 CPUs annually (F_2)	
Modules available	Numerous modules including digital input/output, analog input/output, communications, special modules	
Max. I/O[1] (F_2 /Fx)	120	256
Timers	32	256
Counters	32	256
Internal relays (auxiliaries)	192 (64 battery-backed) 1024 (524 battery-backed)	
Programming terminals	Hand-held, portable graphic programmers and Medoc software for personal computers	
Other ranges	A1S (up to 1024 I/O) and A-series (up to 2048 I/O)	

[1] 'Max. I/O' in this section does not include analog signals.

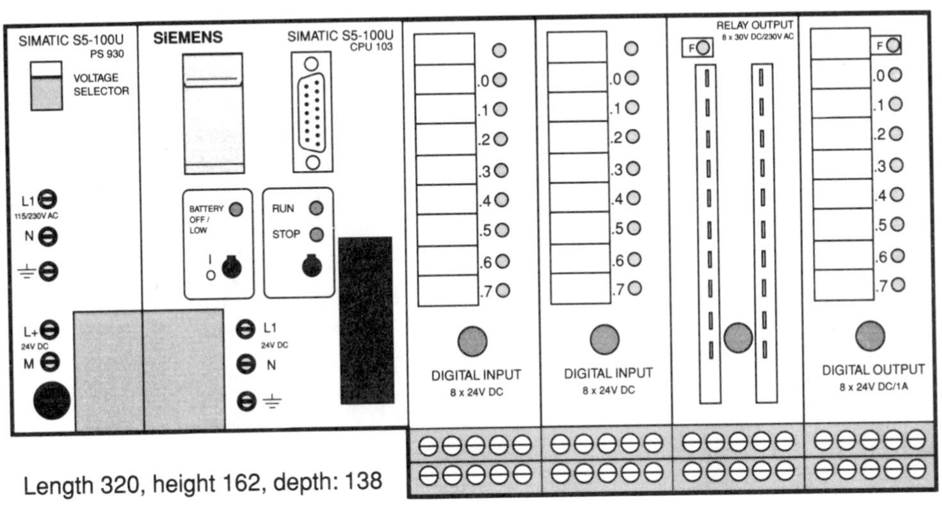

Figure 2.9. The Siemens SIMATIC® S5-100U programmable controller (dimensions in mm).

Company

Name	Siemens
Headquarters	Munich, Germany
Main business activities	Electrical and electronic engineering
Employment world-wide	420 000+

Product

Model	SIMATIC® S5-100U		
Manufactured	Amberg, Germany		
Year introduced	1986		
Production	>100 000 CPUs annually		
Modules available	Numerous modules including CPUs, digital input/output, analog input/output, communications, special modules		
Max I/O (CPU 100, 102, 103)	128	128	256
Timers	16	32	128
Counters	16	32	128
Internal relays (flags)	1024	1024	2048
Programming terminals	Hand-held, portable graphic programmers and S5/S7 software for personal computers		
Other ranges	SIMATIC® S5-90U, S5-95U, S5-115U, S5-135U, S5-155U, S7-200, S7-300; SIMATIC® TI305, TI405 and TI505		

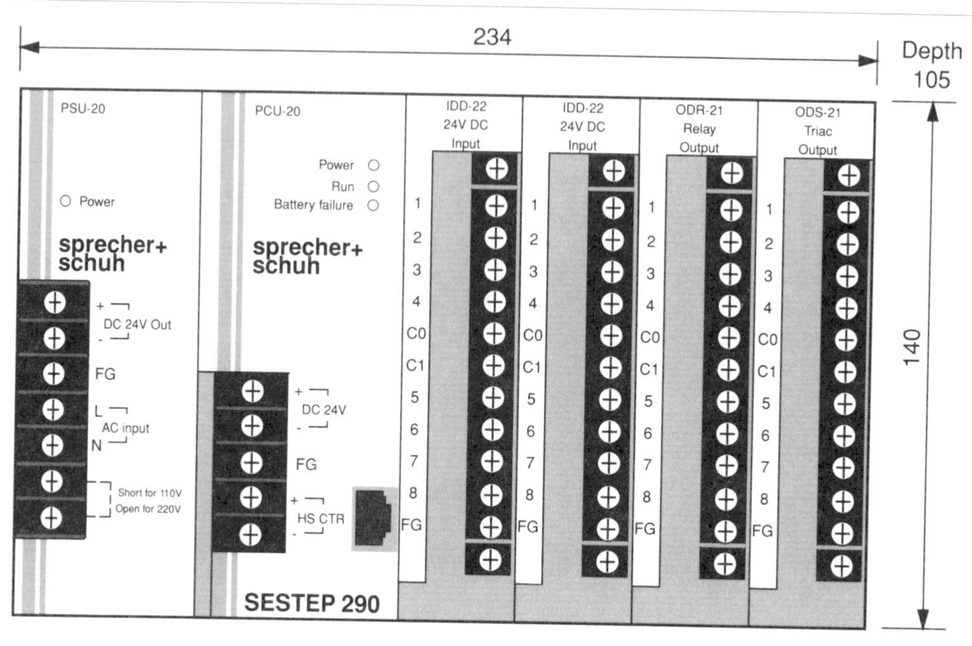

Figure 2.10. The Sprecher+Schuh SESTEP® 290 programmable controller (dimensions in mm).

Company

Name	Sprecher+Schuh
Headquarters	Aarau, Switzerland
Main business activities	Industrial control and automation technology
Employment world-wide	3500+

Product

Model	SESTEP® 290
Manufactured	In several countries
Year introduced	1989
Modules available	Numerous modules including digital input/output, analog input/output, communications, special modules
Max. I/O	128
Timers	64
Counters	64
Internal relays (coils)	256
Programming terminals	Hand-held programmer and SESTEP® software for personal computer
Other ranges	SESTEP® 190, 390, 490, 590 and 690

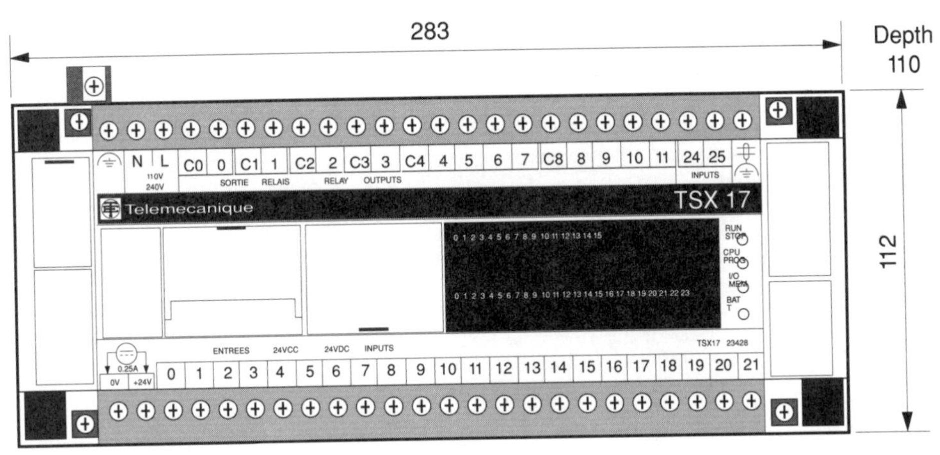

Figure 2.11. The Telemecanique TSX 17_20 programmable controller (dimensions in mm).

Company

Name	Telemecanique
Headquarters	BP 236, 43/45 bd Franklin-Roosevelt, 92504 Rueil-Malmaison, Cedex, France
Main business activities	Automation products and systems
Employment world-wide	14 500+

Product

Model	TSX 17 series	
Manufactured	Sophia Antipolis, Nice, France	
Year introduced	1988	
Modules available	Numerous modules including digital input/output, analog input/output, communications, special modules	
Max. I/O (PL7-1, PL7-2)[2]	120	160
Timers	32	32
Counters	15	32
Internal relays (bits)	256	256
Programming terminals	PL7-1: hand-held or software for personal computer PL7-2: portable graphic terminals or software for computer	
Other ranges	TSX 07 (48 I/O), TSX 47 (512 I/O), TSX 67 (1024 I/O), TSX 87 and TSX 107 (2048 I/O)	

[2] PL7-1 refers to Telemecanique's Boolean (list) language; PL7-2 refers to its Ladder and Grafcet languages.

Chapter 3 Input/Output devices and circuits

3.1 General

In this chapter we examine the input and output devices commonly found in automation systems, their connection to the programmable controller and a selection of the I/O modules available to the purchaser. We deal exclusively with ON–OFF signals (known also as 'binary', 'discrete', or 'digital' signals).

The programmable controller's input and output modules must be compatible with the devices and supplies in use. Four points need to be considered:

- kind of current (a.c. or d.c.)
- rated current
- rated voltage
- polarity.

This information is best obtained from manufacturers' catalogues, data sheets or user manuals. The most commonly used supplies are: 24 V, 48 V, 115 V and 230 V (a.c. or d.c.).

3.2 Input devices

An input signal can be generated by a manual or an automatic device. For example, a pushbutton gives a manual control signal such as START or STOP. The contacts can be make (normally open) or break (normally closed). It is normal practice to obtain a START signal from a make contact and a STOP signal from a break contact (Fig. 3.1).

A selector switch is a manually operated switch having two or more positions (see Fig. 3.1). Typically one contact closes for each selected position, and each contact is connected individually to an input on the programmable controller.

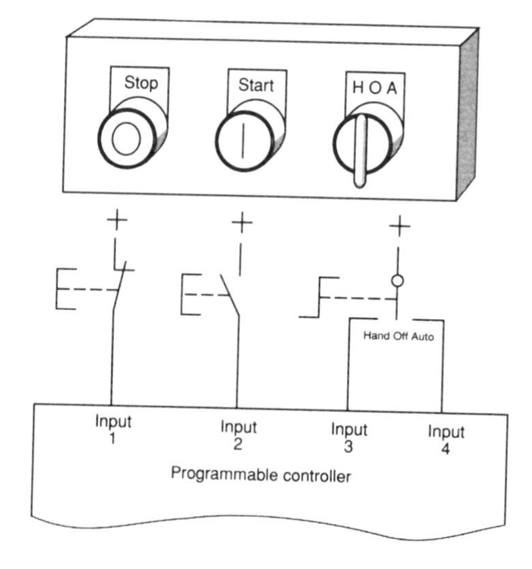

Figure 3.1. Manually generated input signals.

A thumbwheel switch is used to make a numerical selection. The switch itself has four poles (Fig. 3.2); a contact may be open or closed depending on the selected position. For example at position 3, the first and second contacts are closed. The thumbwheel switch can give 10 different selections using only four inputs. By combining two such switches, we can obtain 100 selections using only 8 inputs!

A limit switch is an electromechanical position switch used to detect the passage or travel of a moving part. It can be actuated by cam, roller or lever; it supplies an input signal through its make or break contact (Fig. 3.3).

A thermostat is a heat sensor operating on the expanding liquid or bimetal principle, which acts on a make and/or break contact when a pre-set temperature is reached. Either contact may be used to switch the input signal (Fig. 3.3).

A pressure switch is used for sensing fluid pressure or vacuum. The mechanism incorporates a bellows and snap action springs which act on make and/or break contacts when the set pressure is reached. Either contact may be used for switching the input signal (Fig. 3.4).

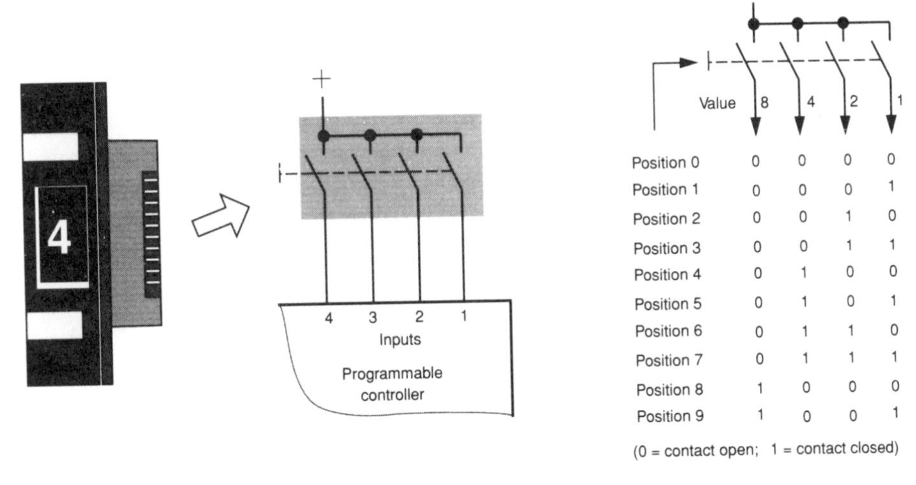

	8	4	2	1
Position 0	0	0	0	0
Position 1	0	0	0	1
Position 2	0	0	1	0
Position 3	0	0	1	1
Position 4	0	1	0	0
Position 5	0	1	0	1
Position 6	0	1	1	0
Position 7	0	1	1	1
Position 8	1	0	0	0
Position 9	1	0	0	1

Figure 3.2. Thumbwheel switch.

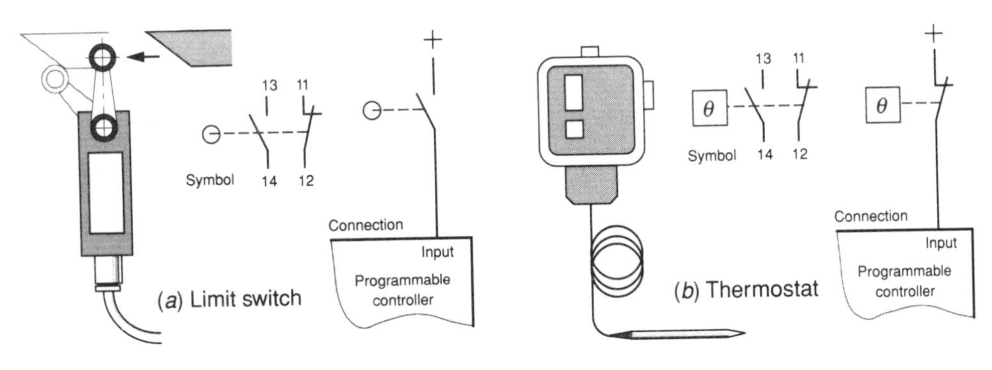

Figure 3.3. Limit switch and thermostat.

A level switch is used for sensing liquid level. Make and/or break contacts are operated by a lever attached to a float mechanism. Either contact may carry the input signal to the programmable controller (Fig. 3.4). Level probes used in conjunction with an amplifier perform a similar function to the level switch. Recently sonar devices have begun to be used for level detection.

A proximity sensor is a solid state device used to detect presence, passage or flow of parts, for positioning, end of travel, or rotation. The inductive type is able to detect a nearby *metal* object because of a change in the magnetic field at its sensing face. The capacitive type is able to detect a nearby *non-metallic* object because of a change in the electric field at its sensing face (see Fig. 3.5).

Two-wire and three-wire sensors are available. The older type of two-wire sensor often has a high off-state current which makes it unsuitable for use with a programmable controller. The three–wire sensor is given its own d.c. supply which is independent of the signal wire; as a result it is much more reliable in programmable controller applications. The three-wire type (Fig. 3.5) can have either PNP or NPN switching. During operation the PNP type switches the positive to the input of the programmable controller; thus the PNP device is used with controllers expecting a positive input signal. The NPN type switches the negative to the input; thus the NPN device is used with controllers expecting a negative input signal.

Photoelectric sensors are solid state devices that use infra-red light for detecting the presence, passage or movement of objects. There are two main types: through-beam and reflex (Fig. 3.6). The through-beam type has a separate transmitter and receiver between which the beam of light passes. The reflex type has a combined transmitter/receiver and is used in conjunction with a reflector. PNP and NPN types are available. The sensor is so positioned that the target object interrupts the light beam as it passes and an input signal is generated accordingly.

The three-wire proximity or photoelectric sensor has a small current demand. In real automation projects, however, large numbers of sensors may be used and the accumulated current could well exceed the rated current of the power supply in the programmable controller. If this problem is foreseen at the planning stage it can be offset by choosing a controller with an adequate power supply. If it is discovered after installation, either the larger power supply must be retrofitted or a new, separate power supply must be provided to serve the sensors.

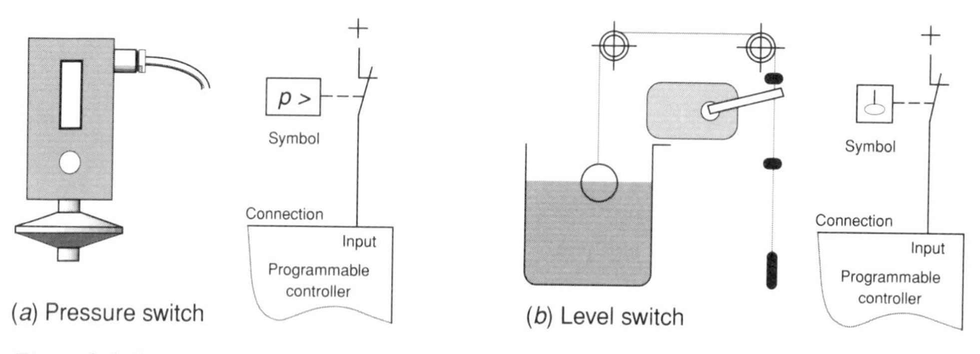

Figure 3.4. Pressure switch and level switch.

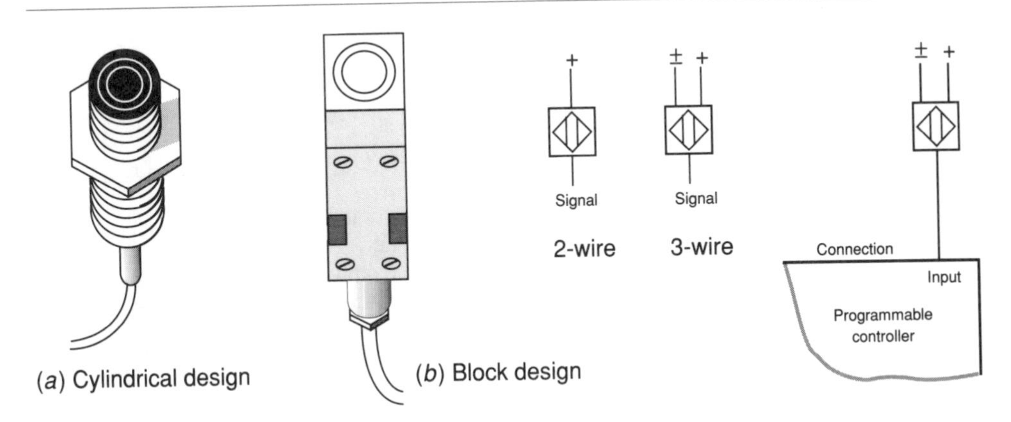

Figure 3.5. Proximity sensors.

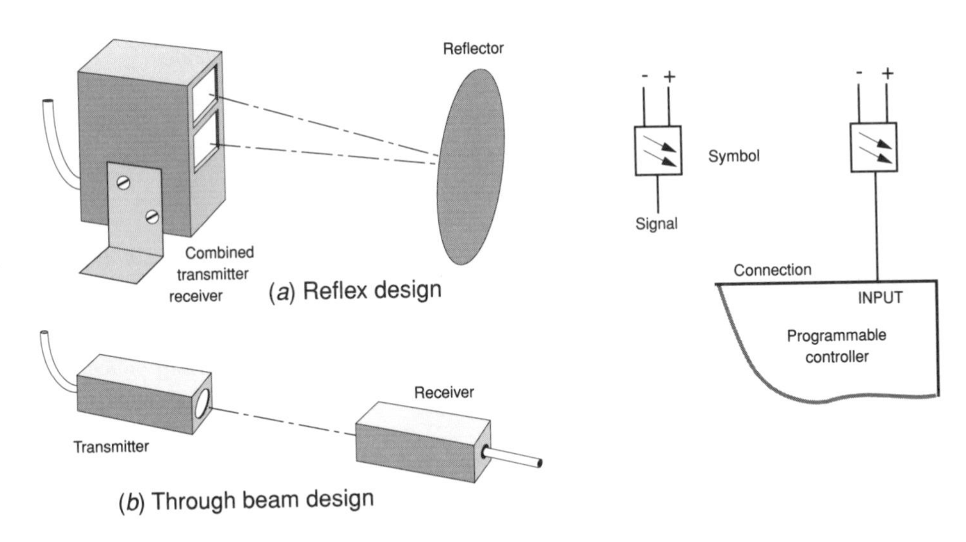

Figure 3.6. Photoelectric sensors.

3.3 Output devices

The contactor is a switching device having a number of contacts which are operated together by its electromagnet (the 'coil'). It is used very extensively for switching motors, heaters, etc.

For automation work we connect the coil of the contactor to an output of the programmable controller. When the output is turned on, the coil is energized and the contactor connects its load to the supply (Fig. 3.7). For small contactors the coil current is modest and can be switched directly by the output of the controller. For large contactors the coil current could exceed the capacity of the output. In this case we let the controller's output drive a relay (a so-called interface relay) which in turn switches the current of the larger contactor coil.

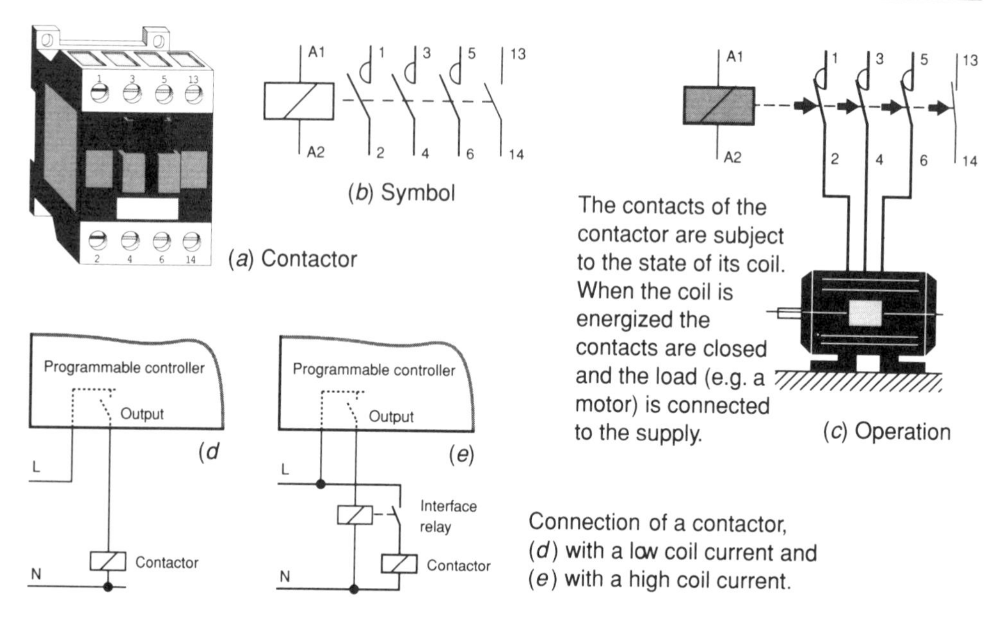

Figure 3.7. Contactor.

Functionally the relay and the contactor are very similar; the main difference lies in the switching capacity of the contacts. A contactor is designed for power switching (currents up to 1600 A) whereas a relay is meant for 'small' currents (≤10 A).

The coil of a relay or a contactor is highly inductive and this means it tends to produce high induced voltages on switch-off. We have to design our output circuits with due regard to this problem (see Section 3.6).

A solenoid valve is another extensively used component. The coil is arranged to open or close the route between two or more 'ports', enabling it to control the flow of liquid or gas. The coil is connected to the controller's output; by this means hydraulic and pneumatic mechanisms can become part of an automation system (Fig. 3.8). Like the relay and contactor, the coil of a valve is highly inductive.

The current drawn by small valves is extremely low (tens of mA) and may be comparable with the leakage current of some output switching devices or their internal protective circuits. This can lead to the valve being energized even when the output is off! In such a case the valve coil may be shunted by a resistor which will 'absorb' the leakage current (Fig. 3.9); alternatively, it can be switched through the contact of an interface relay (Fig. 3.7).

A signal lamp is used to indicate the status of a plant or an operating cycle. It is normally switched directly by the programmable controller as shown in Fig. 3.10. The general-purpose indicator is fitted with an incandescent lamp which presents no difficulty for the output. In recent designs the neon lamp or the high-intensity light-emitting diode (l.e.d.) has gained in popularity. The very low current of this kind of lamp occasionally presents problems (see Section 3.6).

The trend in recent times has been to replace dedicated signal lamps with message displays or operator interfaces. These furnish the operator with a large repertoire of messages in clear text, in the local language and at a single location. There is a drastic reduction in the number of outputs used and a simplification of the wiring.

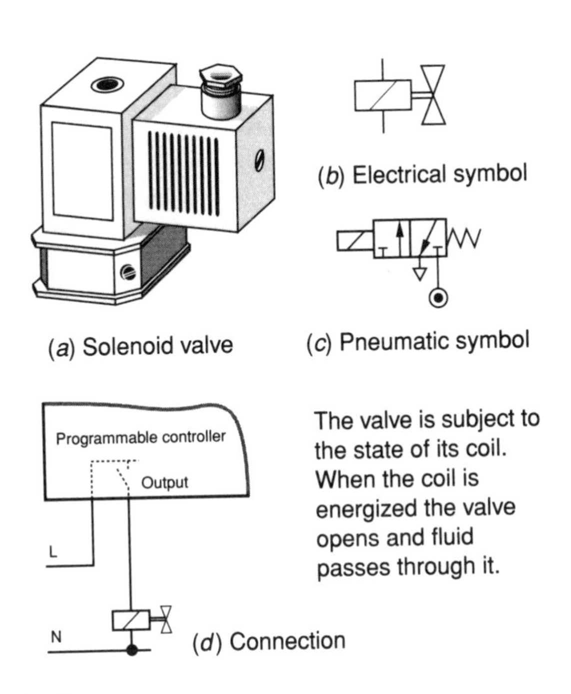

Figure 3.8. Solenoid valve.

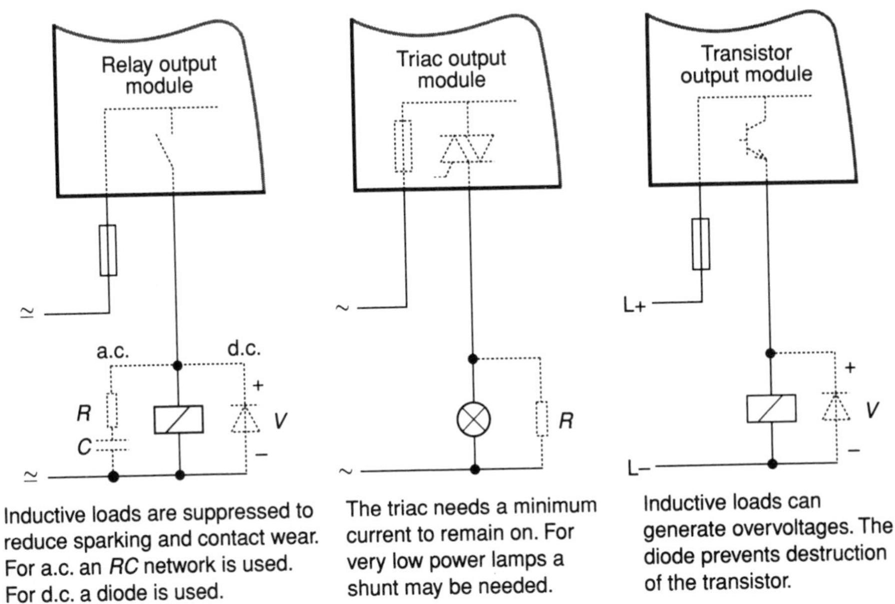

Figure 3.9. Relay, triac and transistor outputs.

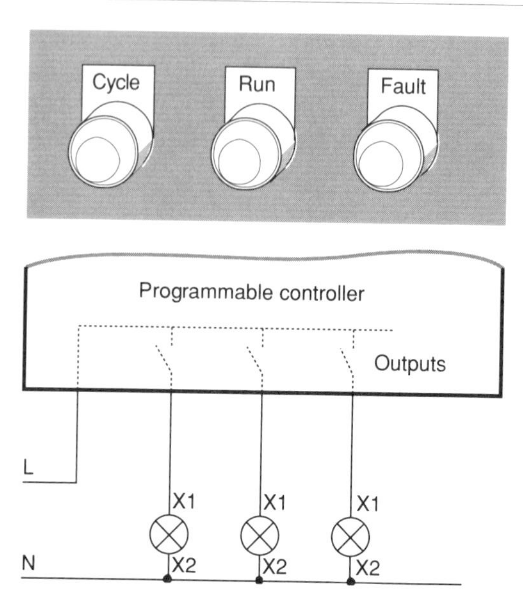

Figure 3.10. Signal lamp.

3.4 Operator interface

In projects of larger scale or complexity, the task of informing the operator takes on a greater significance. Meeting this task with traditional signal lamps can result in large display panels which can be difficult for the operator to survey and which use up great numbers of outputs on the programmable controller.

One solution is the message (text) display, an example of which appears in Fig. 3.11. This device (model UCT-35P from Brodersen Control Systems of Denmark) displays messages of ordinary text, organized in two lines of 20 characters each. Thus the operator may receive

- information DEFROST CYCLE IN PROGRESS
- alarms CONVEYOR MOTOR OVERLOAD
- instructions PRESS START TO ADVANCE
- data WATER TEMP. > 80°C

The messages are loaded in the memory of the display using a personal computer or a keyboard; each message has its own recall number. The display is given its own power supply and is connected to a number of outputs on the programmable controller (nine for maximum exploitation of the model referred to). The controller activates a message by turning on an appropriate pattern of outputs, which the display interprets as a recall number.

Table 3.1 shows how this works using binary code (BCD (binary coded decimal) is also possible). Each output is *weighted*: output 0 is worth 1, output 1 is worth 2, output 2 is worth 4, and so on. The number conveyed to the display is the sum of the values of the outputs that are on. Thus, for example, message 37 (highlighted) requires the following outputs to be on:

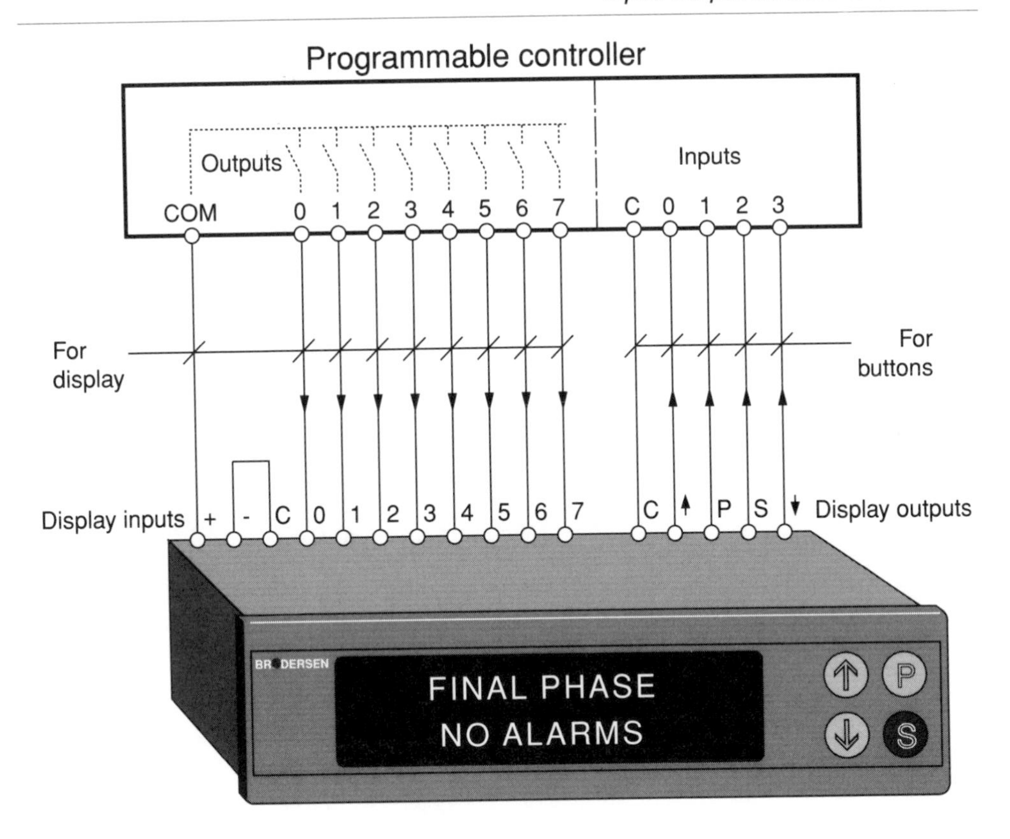

Figure 3.11. Message (text) display.

output 0	worth	1
output 2	worth	4
output 5	worth	32
total worth		37

A message display may be fitted with buttons or keys. In our example there are four keys, which are connected to inputs on the controller. The operator can use these keys to respond to the messages displayed, effectively opening up two-way communication with the programmable controller. A device capable of this kind of service is known as an *operator interface.*

The operator interface, shown in Fig. 3.11, is described in literature as a parallel type and has the following advantages:

- it can be used with any programmable controller, regardless of manufacture
- it requires no communications port on the controller, just inputs and outputs
- it is easy to set up and requires only minor programming work in the controller.

A popular alternative is the serial-type operator interface, shown in Fig. 3.12, for which inputs and outputs are not necessary. Instead, dialogue with the controller takes place via

Table 3.1. Binary representation of decimal values

Output signals								
7 (128)	6 (64)	5 (32)	4 (16)	3 (8)	2 (4)	1 (2)	0 (1)	Decimal value
0	0	0	0	0	0	0	0	0
0	0	0	0	0	0	0	1	1
0	0	0	0	0	0	1	0	2
0	0	0	0	0	0	1	1	3
0	0	0	0	0	1	0	0	4
0	0	0	0	0	1	0	1	5
0	0	0	0	0	1	1	0	6
0	0	0	0	0	1	1	1	7
0	0	0	0	1	0	0	0	8
0	0	0	0	1	0	0	1	9
0	0	0	0	1	0	1	0	10
0	0	0	1	1	0	0	0	24
0	0	1	0	0	1	0	1	37
0	0	1	1	0	1	0	0	52

the programming or other port. This demands full compatibility between the controller and interface, which is most easily assured by choosing both from the same manufacturer. The advantages are that

- inputs and outputs are not used and are available for other purposes
- all parameters in the control program are fully accessible
- simple, low-cost versions are available from each manufacturer.

3.5 Input modules

A programmable controller input module is selected to suit the control voltage and the input devices proposed to be connected to it. Figure 3.13 illustrates the content and function of a d.c. input module, in a simplified form.

Each incoming input signal passes through a status l.e.d. (for visual indication on the module), a voltage dropping resistor and an opto-isolator (which consists of an l.e.d. and photo-transistor). In an a.c. module it also passes through a rectifier. The objective is to provide the microprocessor with a signal it can handle—5 V d.c., free of interference

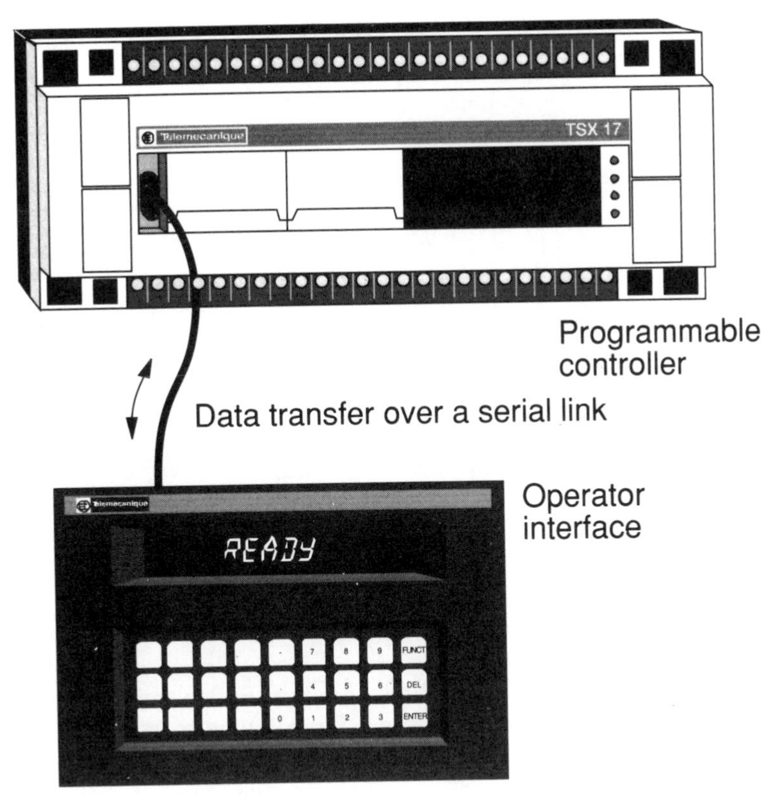

Figure 3.12. Serial operator interface.

('noise') and electrically isolated. Each input module will accommodate a number of input signals, e.g. eight for the Siemens SIMATIC® S5-100U and the Sprecher+Schuh SESTEP® 290.

The relevant catalogue should be consulted before choosing the controller or input module, for example

1. we have 21 inputs for 24 V d.c. and we want a Mitsubishi controller: we choose an F_2-40MR-ES or F_2-40MS-ES or F_2-40MR-DS (all accept 24 inputs at 24 V d.c.; they differ only in respect of outputs or power supply)
2. we have 14 inputs for 230 V a.c. and we want a Siemens controller: we choose two modules 6ES5-8MD11 (each accepts 8 × 230 V a.c. inputs).

3.6 Output modules

An output module of the programmable controller transfers the output signals from the microprocessor at very low voltage to the output terminals of the controller at plant voltage, typically 115/230 V a.c. Once again, opto-isolators provide electrical isolation and limit the effects of interference.

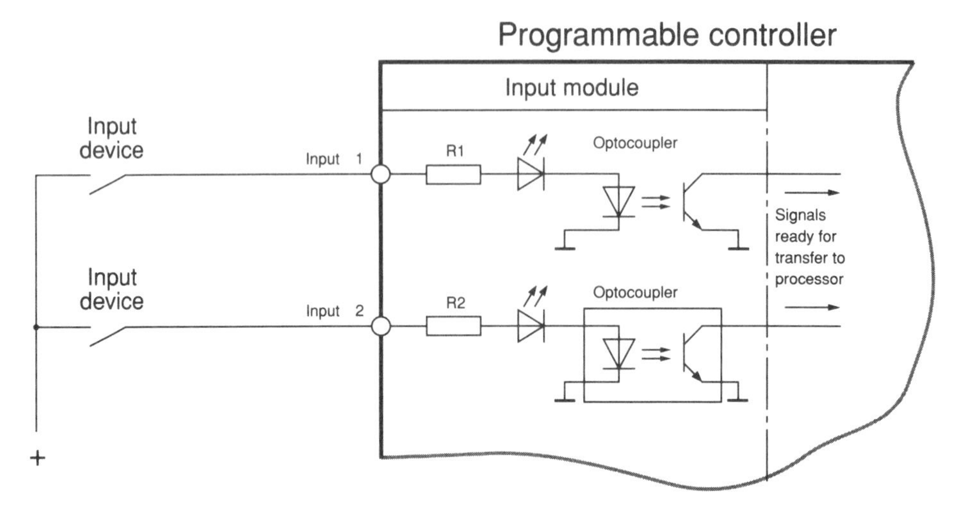

Figure 3.13. Direct current input module.

Figure 3.14 illustrates how a relay output module functions. Each outgoing signal is routed through an opto-coupler for isolation, then an l.e.d. for indication on the module, and finally turns on the coil of a sub-miniature relay. The contact of this relay connects the supply to the output device. In other modules the place of the relay is taken by a triac or a transistor (see Fig. 3.9 also).

The relay output

- is suitable for a.c. and d.c. switching
- provides real separation when the contact is open
- can withstand high surge currents and severe voltage transients
- is subject to mechanical failure and contact erosion.

The triac output

- is strictly for a.c. switching
- is silent, has no moving parts and does not suffer from contact wear
- is easily destroyed by overcurrent.

The transistor output

- is strictly for d.c. switching
- is silent, has no moving parts and does not suffer from contact wear
- is capable of switching at very high speed
- can be destroyed by overcurrent and high reverse voltage.

In general the triac output is the most prone to damage by overcurrent, and a fuse is nearly always incorporated in the module to protect each output or group of outputs. This type of fuse is specially adapted for semiconductor protection and must be used and replaced in strict accordance with the manufacturer's advice. For a triac to remain conductive, the

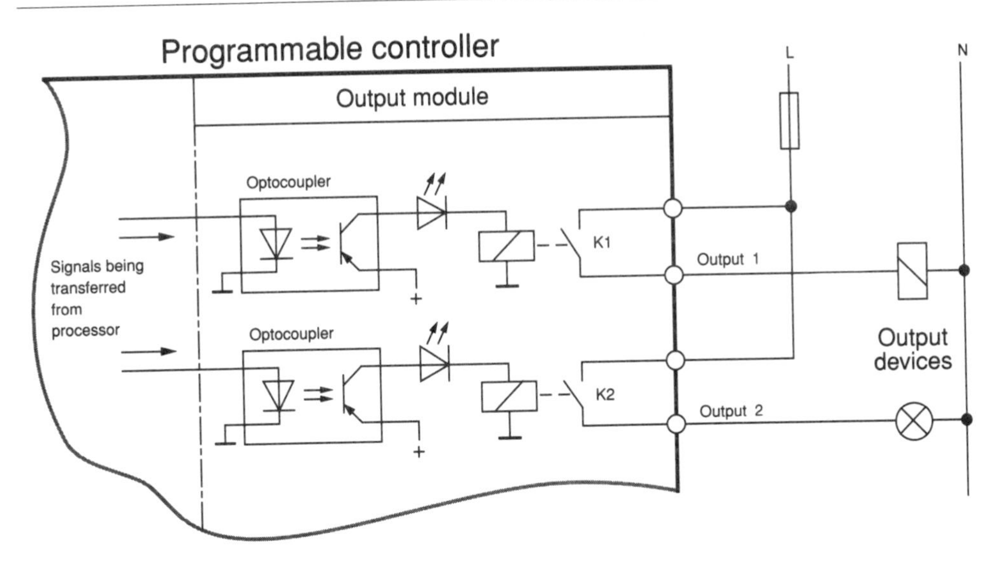

Figure 3.14. Relay output module.

current must exceed a minimum value known as the 'holding current'. Some low power loads may need a shunt resistor to meet this requirement (see Fig. 3.9).

The transistor output is also prone to overcurrent damage, but nowadays many transistor modules come equipped with built-in electronic protection (so-called short-circuit-proof outputs). An external fuse or circuit breaker may none the less be fitted to protect the power supply and the wiring.

Relay outputs are the most robust and protection is straightforward: each output or group of outputs is protected with a fuse or circuit-breaker of suitable rating, fitted externally.

Once again we need to refer to a catalogue to choose the correct controller or output module, for example

1. we have 19 × 230 V a.c. outputs and we want a Sprecher+Schuh controller: we choose three modules ODR-21 or ODS-21 (8 × 250 V a.c. outputs; ODR-21 has relays while ODS-21 has triacs)
2. we have 18 × 115 V a.c. outputs and we want a Telemecanique controller: we choose a controller TSX 172 3444E or a TSX 172 3428E (12 × 115 V relay outputs) and extension module TSX DSF 635 (6 × 115 V relay outputs).

Relay, triac and transistor outputs can be damaged or even destroyed by overvoltage. Although protective circuits are included in the output module by the manufacturer, it is better to eliminate the problem at source.

The most common sources are the coils of relays, contactors and solenoid valves, which are highly inductive. Overvoltages from these sources can be suppressed by fitting 'snubbers' as shown in Fig. 3.14. For a.c. circuits either an *RC* network or a varistor is used; for d.c. circuits a reverse-biased diode is used.

3.7 Exercises for Chapter 3 (Solutions in Appendix 1)

3.1. What type of output module is best suited to an application involving the frequent switching of d.c. loads?

3.2. Why should we *not* connect three stop switches in series and return one input from them to the programmable controller?

3.3. State the minimum number of inputs required for making four selections.

3.4. Why is a separate interface relay sometimes used?

3.5. State three advantages of the operator interface.

4.1 General

The program is the 'set of rules' according to which the programmable controller behaves. Not surprisingly, the programmable controller can do nothing without a program. In the first instance the program is created with a *programming terminal*, which has other very important uses such as

- monitoring the program, to help with fault-finding
- modifying the program, to accommodate changes in the plant
- transferring the program to another controller or memory
- documenting the program with, for example, a printer.

The programming terminal is usually detachable and can be moved from one controller to the next without disturbing operations.

One of the advantages often cited for the programmable controller is that it can be programmed by non-specialists. This has become possible with the availability of the intelligent programming terminal. The terminal performs the crucial task of translating the user program (stated in suitable engineering terms) to machine language, the form in which it resides in the memory and is treated by the processor (see Fig. 4.1).

There are several languages for formulating the program. The recently published IEC 1131-3 (*Programmable controllers—Part 3: Programming languages*) recognizes four:

1. Instruction List	IL	text-based
2. Structured Text	ST	text-based
3. Ladder Diagram	LD	graphical
4. Function Block Diagram	FBD	graphical

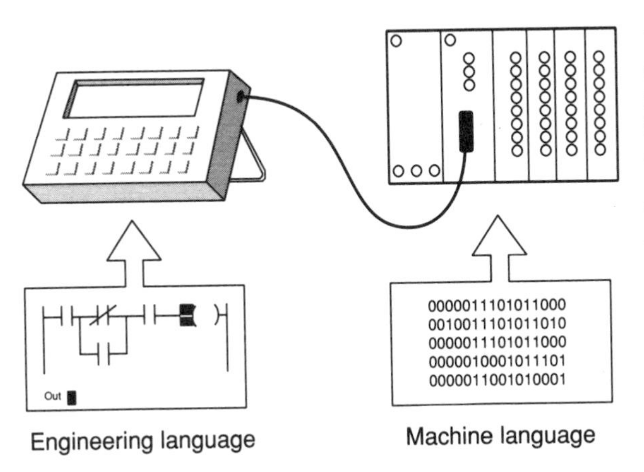

The user needs to state his/her requirements in engineering terms, for example, in list form or in a graphical form such as a ladder diagram.

The controller can work only with machine language, a pure binary code which is difficult to manipulate.

The programming terminal acts as translator between the two languages.

Figure 4.1. The role of the programming terminal.

Step	Instruction	Address
000	A	I0.0
002	AN	I0.1
004	A	Q2.4
006	O.	F1.7
008	A	T3
00A	A	I0.5
00C	O.	C10
00E	AN	Q3.3
010	=	Q2.0

(*a*) Sample of a list program

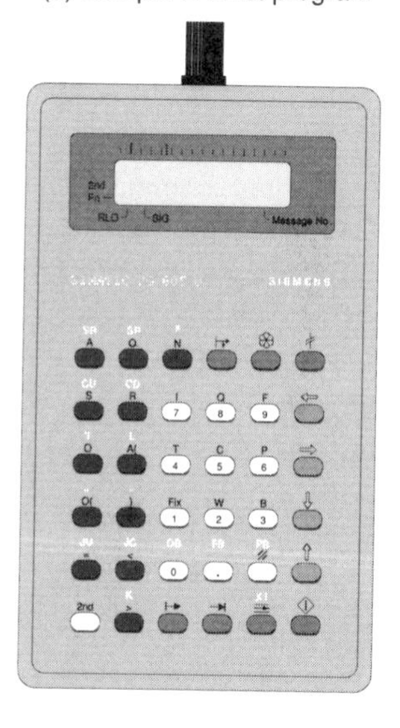

(*b*) A Siemens SIMATIC® PG 605 U hand-held list programming terminal

Figure 4.2. Example of a list program and a compact list programming terminal.

Sequencial Function Chart (SFC) is listed in the standard as an 'organizational element' but many regard it informally as a fifth language.

For the newcomer, the most accessible languages are instruction list and ladder diagram (explained in this chapter) and sequential function chart (explained in Chapter 6).

The list form is relatively close to machine language and can be produced with a modest programming terminal of compact design. It takes the form of logical statements that are assembled to match the control functions required. See Fig. 4.2 for examples of a list program and a programming terminal.

The ladder method is a graphical representation of the program. Symbols depicting the various elements of the program are assembled with horizontal and vertical links to create the required control functions. Ladder programming involves considerable translation effort

(*a*) Example of a graphic programming terminal: Telemecanique TSX 407 1

(*b*) Sample of a ladder program created with the terminal

Figure 4.3. Ladder program and graphic programming terminal.

on the part of the terminal, which is more intelligent, powerful and costly than one designed for list programming only (see Fig. 4.3). A printer interface is normally supplied in terminals of this kind; this facilitates the connection of a printer for a paper copy of the program.

Most manufacturers now offer software packages that allow a standard personal computer to be used as a programming terminal. The advantages include enhanced functionality and performance, against which must be balanced the computer and software costs.

4.2 Defining the problem

Before a program can be written for any controller the user must have a clear picture of the control requirements. This is the *specification.* It can be communicated verbally or in writing. It can also be presented in the form of a circuit diagram or a logic diagram in accordance with IEC 113 or a sequence diagram in accordance with IEC 848 (see Fig. 4.4).

The circuit diagram has been the standard method of representation in control systems from very early times and continues to be widely used in the electrical industry. The newer logic diagram, which is technology-independent, may appeal more to those from electronic or pneumatic backgrounds. If the circuit or logic diagram is used as a model, *the program will focus on correctly stating the conditions for turning on each of the outputs.*

The sequence diagram is technology-independent and vendor-independent and has gained in popularity in recent years. A program modelled on this kind of diagram *will focus on correctly stating the order of the steps in the controlled process.*

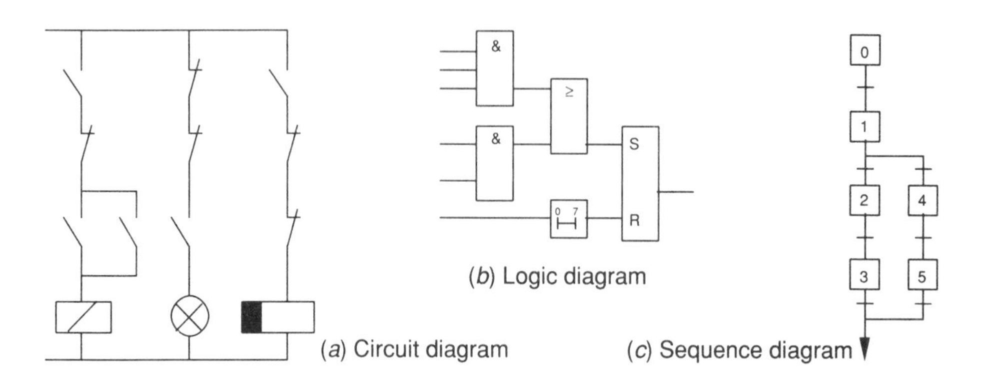

(*b*) Logic diagram

(*a*) Circuit diagram

(*c*) Sequence diagram

Figure 4.4. Different starting points for devising a program.

4.3 List programming

The principles of list programming are illustrated in Fig. 4.5. This example has been written for the Sprecher+Schuh controller but can be adapted for the other controllers. The control requirement is that the lamp H1 is to be turned on only when all these conditions are met

- S1 is closed
- S2 is closed
- S3 is open.

We cannot devise a program in terms of switches being 'open' or 'closed'; instead we have to stipulate the *signals* to be obtained from the inputs to which the switches are connected. Thus we can restate our conditions as follows

- S1 closed = signal at input X001
- S2 closed = signal at input X002
- S3 open = no signal at input X003.

The STR instruction is used to start each new string of instructions in the program; the AND instruction effectively ensures that all the required signals must be in place and the OUT instruction is the call to turn on the output. The instructions used on the other controllers are similar, and in some cases identical. To visualize the processor's treatment of this program in the RUN mode, consider the illustrations in Fig. 4.6.

(a) All switches are open, i.e. no signals at any of the three inputs; the condition at line 002 is true—there is no signal at X003.

(b) S1 has been closed; this brings a signal to X001 and the condition at line 000 is therefore true.

(c) S2 is now closed as well; this brings a signal into X002 and makes the condition at line 001 true.

The processor now 'sees' that every condition required for turning on output Y001 has been met, and turns it on.

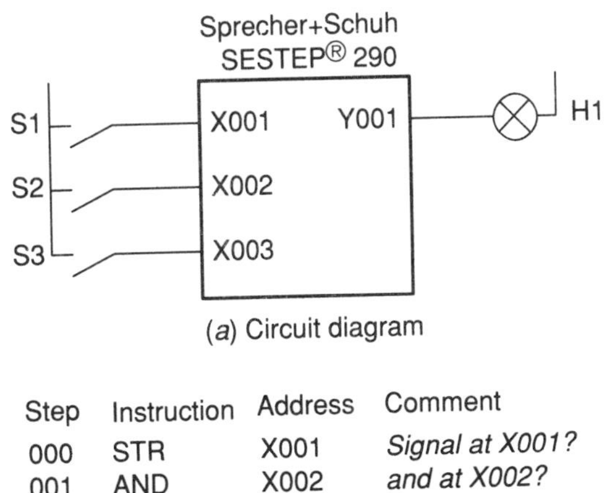

(*a*) Circuit diagram

Step	Instruction	Address	Comment
000	STR	X001	*Signal at X001?*
001	AND	X002	*and at X002?*
002	AND NOT	X003	*and not at X003?*
003	OUT	Y001	*then turn on Y001.*

(*b*) Program in list form

Figure 4.5. Example of a list program on a Sprecher+Schuh SESTEP® 290.

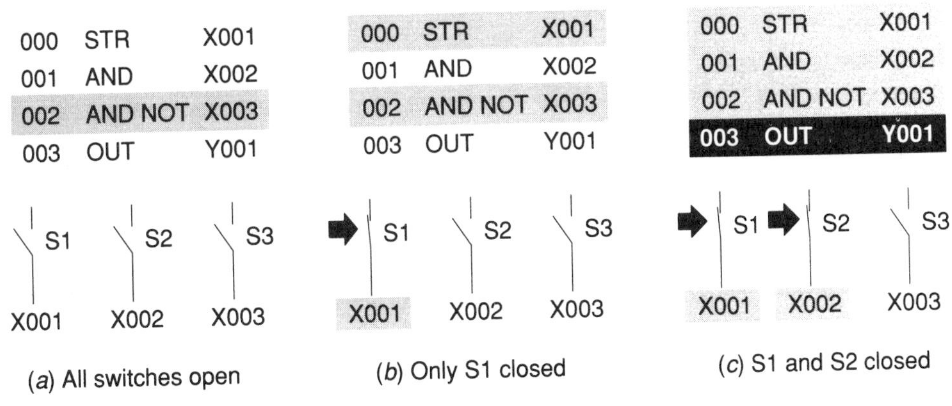

Figure 4.6. Visualizing the execution of a list program.

4.4 Ladder programming

Ladder programming requires the user to 'draw' a diagram on the screen of the programming terminal. The example shown in Fig. 4.7 was written for the Mitsubishi F_2-40 but the principles are the same for the other controllers. We have set the same conditions as for the last example, i.e. to turn on the lamp:

- S1 must be closed
- S2 must be closed
- S3 must be open.

Once again our program must be devised in terms of *signals*:

- S1 closed = signal at X401 =⊣ ⊢X401
- S2 closed = signal at X402 =⊣ ⊢X402
- S3 open = no signal at X403 =⊣/⊢X403

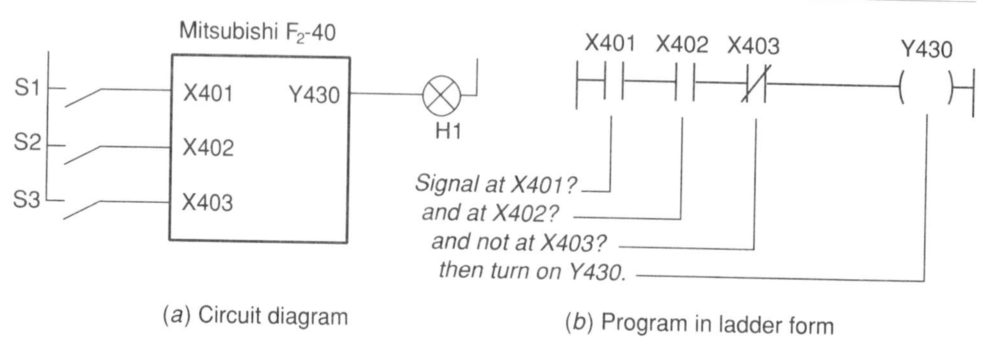

Figure 4.7. Example of a ladder program on the Mitsubishi F_2-40.

We can visualize the processor's reaction to this program in the manner suggested by the illustration in Fig. 4.8.

(a) All switches are open, i.e. no signals at any of the three inputs; however, the condition pertaining to X403 is true and it is therefore shown in bold.

(b) S1 has been closed; this brings a signal to X401 making that condition true.

(c) S2 is now closed as well; this brings a signal into X402 and makes the final condition true.

Since there is a complete line of 'true conditions' stretching from the extreme left to the output symbol '—( )—', output Y430 is turned on. In learning to program using ladder, it may be helpful to keep the following points in mind.

- There are two distinct sections to a ladder diagram. In the right-hand area we place the symbols for actions—like turning on outputs, relays, etc. In the left-hand area we place the symbols for the *conditions* we wish to attach to the actions (Figure 4.9).
- The symbols '—| |—' and '—|/|—' represent signals to be checked by the controller —they *do not represent the contacts of the input devices*. The student is cautioned against drawing comparisons!

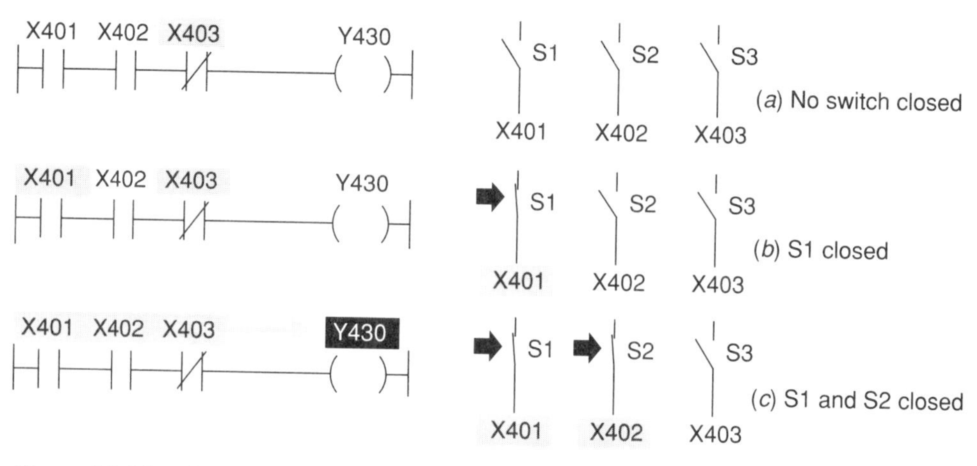

Figure 4.8. Visualizing the execution of a ladder program.

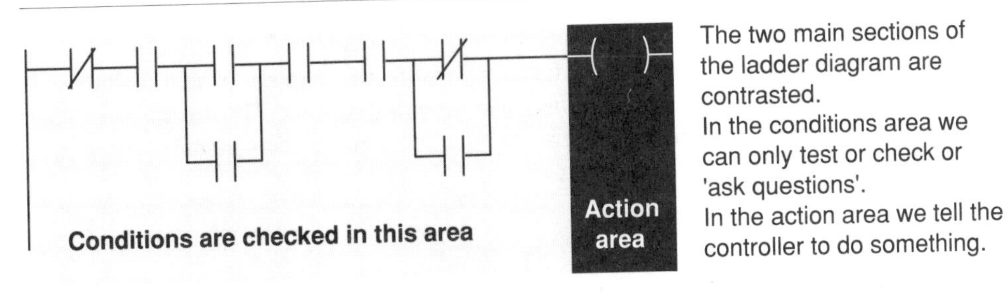

Figure 4.9. Two distinct areas of the ladder diagram.

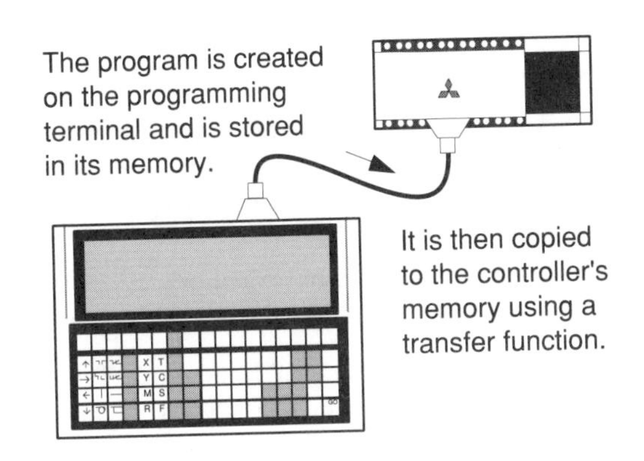

Figure 4.10. Transferring a program.

4.5 Entering the program

The program is entered with the controller in the STOP mode. Some controllers facilitate programming in the RUN mode, but this is potentially dangerous and should be undertaken only when absolutely necessary, and even then after the most thorough consideration.

Entering a program in list form is straightforward, and there are similarities of approach among the manufacturers. Once the terminal has been prepared to accept a new program the various statements are keyed in directly, in the correct order and observing the correct syntax and I/O addresses for the particular controller. It is generally necessary to use the ENTER key (or the GO or WRITE key) to validate each statement. Transfer of the program to the controller may not be automatic; on terminals equipped with their own memories a transfer command is usually called for (see Fig. 4.10).

Entering a program in ladder form generally calls for more planning and care, and there is less uniformity of approach among the manufacturers. Possible differences include

- use of dedicated keys or soft keys for placing symbols
- the maximum number of 'rungs' that can be displayed at any time
- the maximum number of outputs that can be displayed at any time
- acceptable 'circuit' designs
- use of block-type symbols
- program conversion and transfer procedures.

Many of these points are highlighted in Figs 4.11–4.14; note however that these illustrations are no substitute for the manufacturers' instruction manuals, which should always be available to the person performing the programming work. The use of pre-printed ladder preparation sheets (obtainable from some manufacturers) is recommended for a rational and error-free entry of the program.

4.6 Program storage

For any given project, two copies of the program are usually kept. One resides in the memory of the programmable controller; this is the working copy. The other is a backup copy, kept as a precaution against the corruption or loss of the first.

The memory of the programmable controller can be one of three kinds:

- random access memory (RAM), which is freely programmable at any time, but has the unfortunate drawback of clearing itself when it loses power
- programmable read-only memory (PROM), which is secure against loss of power but cannot be programmed freely
- electrically erasable programmable read-only memory (EEPROM or EAPROM), which is also secure against loss of power and can be reprogrammed fairly easily.

The programmable controller is delivered with a RAM memory. This allows the user to develop the new program and to make any changes necessary to perfect it (in a real project several modifications and revisions are typical). It has its own battery which, to some extent, offsets the worries of losing the program. The battery is normally a long-life lithium type and its voltage is monitored by the controller. A signal lamp on the front of the

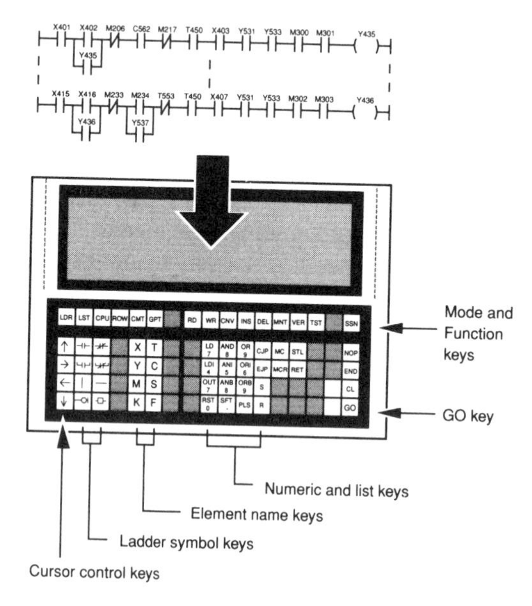

Programming features

1. Up to seven 'rungs', each with up to 11 elements and the output, can be viewed on the LCD display.

2. Eight dedicated ladder symbol keys and four cursor keys.

3. Block functions are not used in the middle of a 'rungs'.

4. Upward branching is not allowed; branching involving two or more outputs is restricted.

5. The terminal has its own memory. The ladder program must first be converted with the key sequence CNV GO and then transferred to the controller by using the key sequence CPU WR GO.

Figure 4.11. The Mitsubishi GP80 graphic programming terminal.

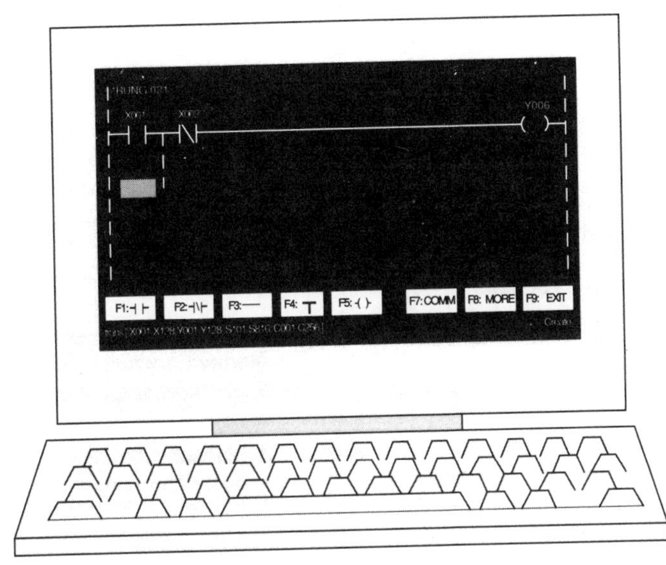

Programming features
(with a personal computer as programming terminal)

1. One screen can display seven 'rungs' each with up to eight elements and an output. 2. Symbols are placed with the soft keys F1 to F8. The standard keyboard supports entry of I/O names, values, etc.
3. Block-type functions are used for timers, etc.
4. No upward branching; branching involving more than one output is limited.
5. A communication function allows transfer of the program between the terminal and controller.

Figure 4.12. A personal computer acting as a programming terminal for the Sprecher+Schuh SESTEP® 290 programmable controller.

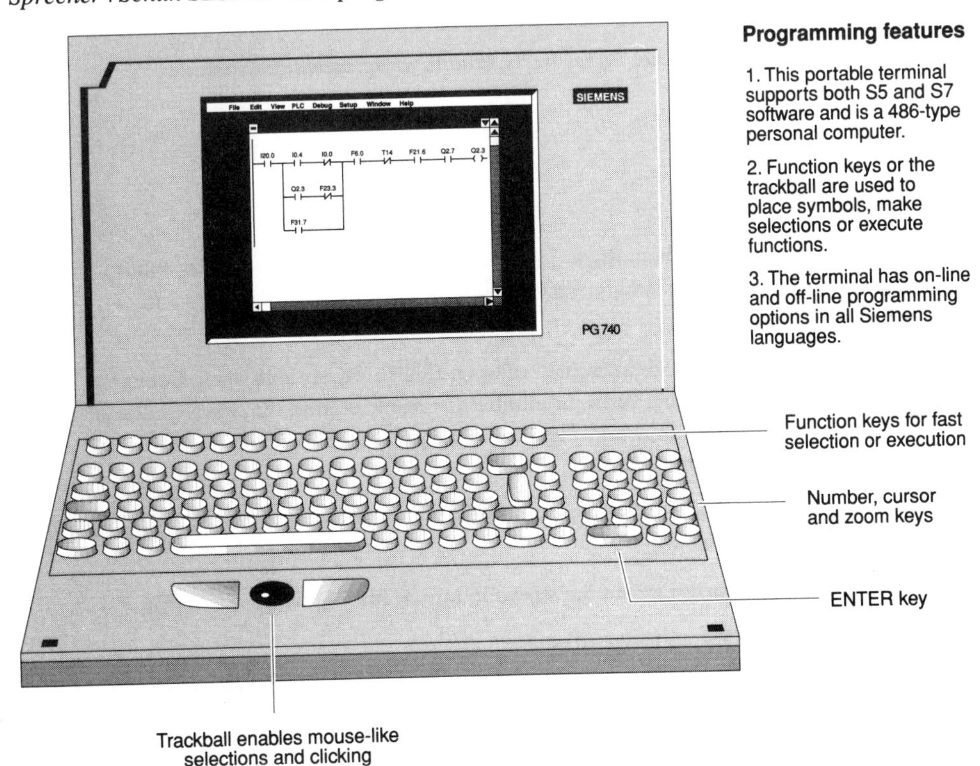

Figure 4.13. The Siemens PG740 graphic programming terminal.

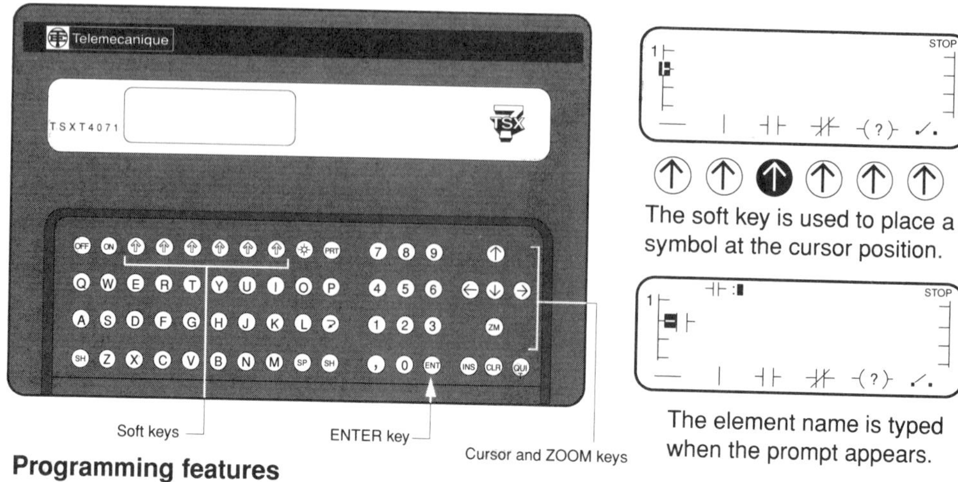

Programming features

1. The ladder is broken up into *networks*, each with a label number. One network with up to four 'rungs' fills the LCD screen. Each rung can have up to nine elements and an output.
2. There are six soft keys for symbol placement and four cursor keys.
3. Block-type functions are used extensively; the ZOOM key gives access to a block.
4. Within each network, lines can move up or down; branches can be 'connected' freely.
5. This terminal has its own memory. It can be programmed off-line (local mode) or on-line (connected mode). The ENTER key is used to effect on-line transfer. The TRANSFER function is used to transfer a program that was created off-line.

Figure 4.14. The Telemecanique TSX T407 1 graphic programming terminal.

controller illuminates when the voltage is low to warn the user to arrange for battery replacement (Fig. 4.15). This is one of the few maintenance tasks necessary with a programmable controller.

Once a program is proven it can be copied onto an EEPROM or other secure memory. This memory, in cartridge or module form, is suitable for insertion into the memory slot of the programmable controller as shown in Fig. 4.16. Once fitted, it may be accepted preferentially as the program to be run.

All work relating to insertion and removal of memory cartridges should be undertaken only when the power is switched off.

The backup copy of the program can be stored in any of the following ways (Fig. 4.17):

(a) PROM or EEPROM cartridge (discussed above)

(b) audio cassette tape, which is very economical but also very slow (for both storing and retrieving programs)

(c) computer disk (fixed or floppy)—this is an option only if the program has been developed or retrieved using the manufacturer's proprietary software.

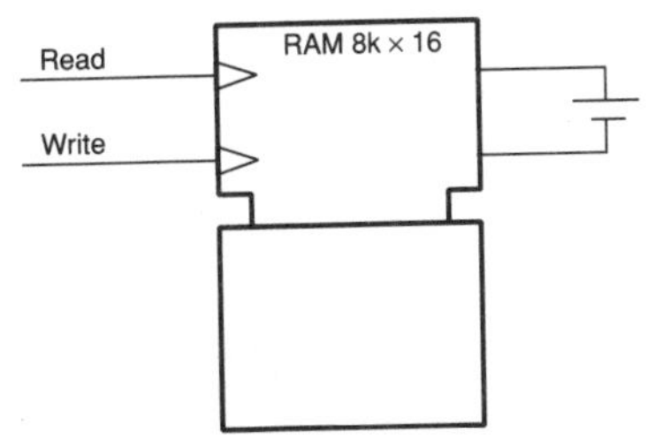

(*a*) RAM symbol with battery support

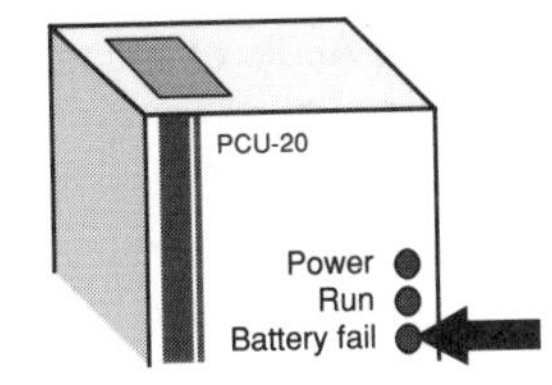

(*b*) Low battery warning lamp

Figure 4.15. RAM memory.

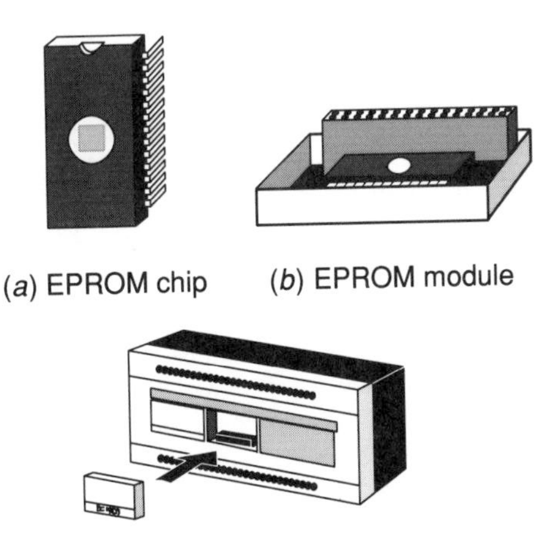

(*c*) The EPROM or EEPROM module can be plugged into the memory socket on the controller, with the power off

Figure 4.16. Memory cartridge.

4.7 Worked examples

The worked examples on the following pages are designed to explore the basic control functions of the programmable controller. The scheme is such that the same connection diagrams, shown in Fig. 4.18, apply in every case. We have done this deliberately in order to emphasize the fact that no wiring changes are needed to accomplish a change of control

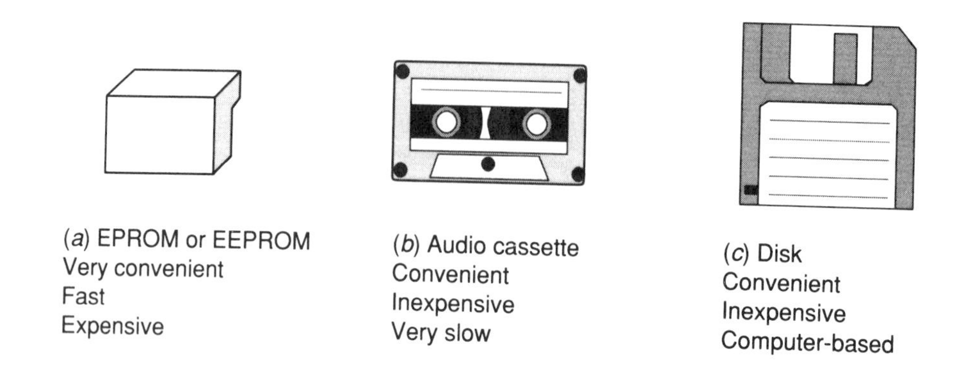

Figure 4.17. Methods of storing backup copies.

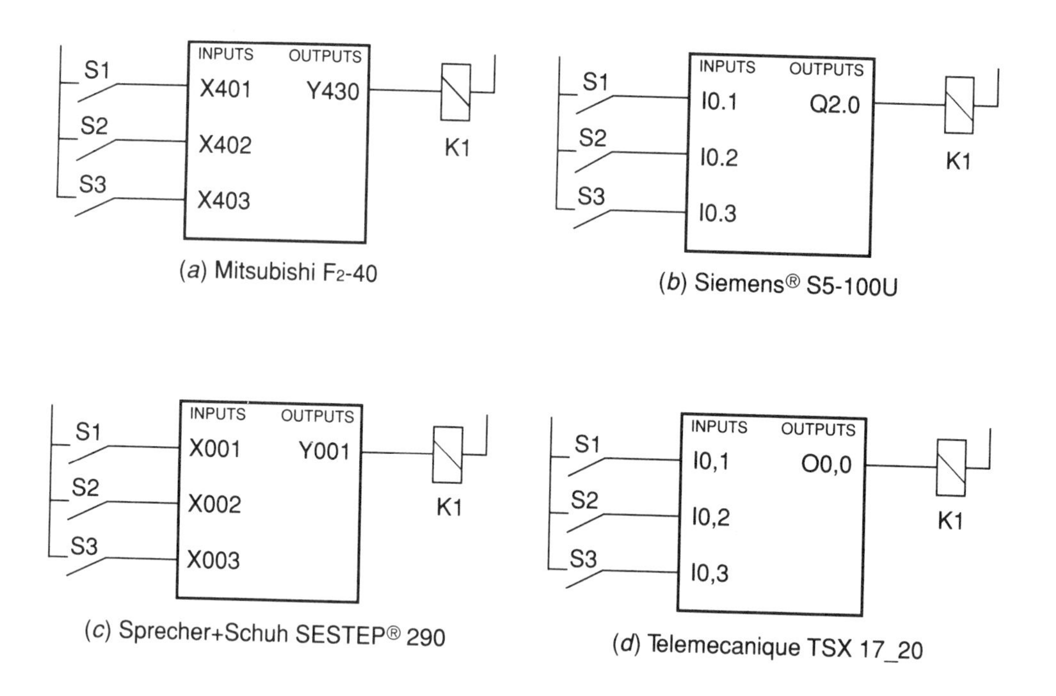

Figure 4.18. Connection diagrams for the exercises.

behaviour in a programmable controller. This also means, however, that the memory must be cleared before starting on each new exercise. The correct procedures for doing this will be found in the manufacturer's user manuals.

Because they call for clearing of the memory, these examples should not be undertaken on programmable controllers that are in service.

Each example follows the same format: a circuit diagram and a logic diagram are offered as a model for program development. The solution is then presented in list and ladder language for each of the four controllers.

4.7.1 *Worked example A*

We require a program that turns on K1 when S1 and S2 and S3 are closed. This corresponds to the series electrical circuit and to the logic AND function, both shown in Fig. 4.19.

The conditions:

- a signal from S1
- AND a signal from S2
- AND a signal from S3
- turns on K1.

The list programs almost restate these conditions. Apart from I/O names, the four programs

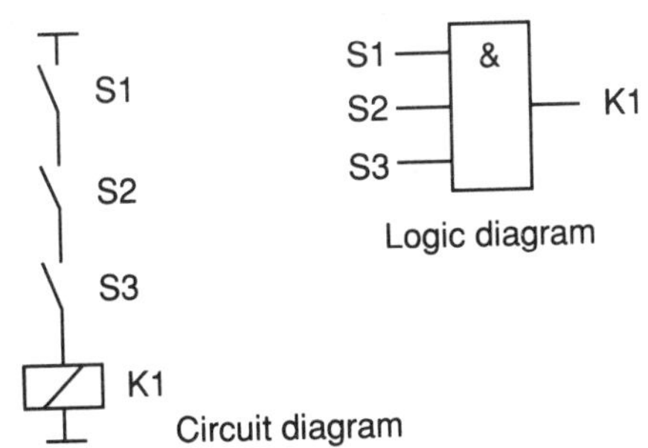

Ladder diagram — List

X401 X402 X403 Y430 — LD X401 / AND X402 / AND X403 / OUT Y430

Mitsubishi F2-40

Ladder diagram — List

X001 X002 X003 Y001 — STR X001 / AND X002 / AND X003 / OUT Y001

Sprecher+Schuh SESTEP® 290

I0.1 I0.2 I0.3 Q2.0 — A I 0.1 / A I 0.2 / A I 0.3 / = Q2.0

Siemens SIMATIC® S5-100U

I0,1 I0,2 I0,3 O0,0 — L I0,1 / A I0,2 / A I0,3 / = O0,0

Telemecanique TSX 17_20

Figure 4.19. The series circuit (AND function).

are distinguished by

- the first instruction (**LD**, **A**, **STR**, **L**)
- the AND instruction (**AND**, **A**) and
- the instruction for turning on the output (**OUT**, =).

It is self-evident that the ladder diagrams are identical apart from the I/O names.

4.7.2 Worked example B

We require a program that turns on K1 when S1 or S2 or S3 is closed. This corresponds to the parallel electrical circuit and to the logic OR function, both shown in Fig. 4.20.

The conditions:

- a signal from S1
- OR a signal from S2
- OR a signal from S3
- turns on K1.

The list programs restate these conditions. The OR instruction is slightly different in the four programs: Mitsubishi and Sprecher+Schuh use **OR**; for Siemens it is **O.** (the period is important!) and Telemecanique uses **O** (without a period).

The ladder diagrams are identical apart from the I/O names.

4.7.3 Worked example C

We require a program that turns on K1 when (S1 or S2) and S3 are closed. This corresponds to the parallel/series electrical circuit and to the logic OR before AND function, both shown in Fig. 4.21.

The conditions:

- a signal from S1 OR S2 (taken first)
- AND a signal from S3
- turns on K1.

The list programs reflect these conditions. Apart from I/O names, the differences between the four programs are very minor. Note that there is no difficulty in mixing AND and OR instructions in the way indicated; we just need to be careful with the order of the ANDs and ORs to be sure we get the required result .

Once again, the ladder diagrams are identical except for the I/O names.

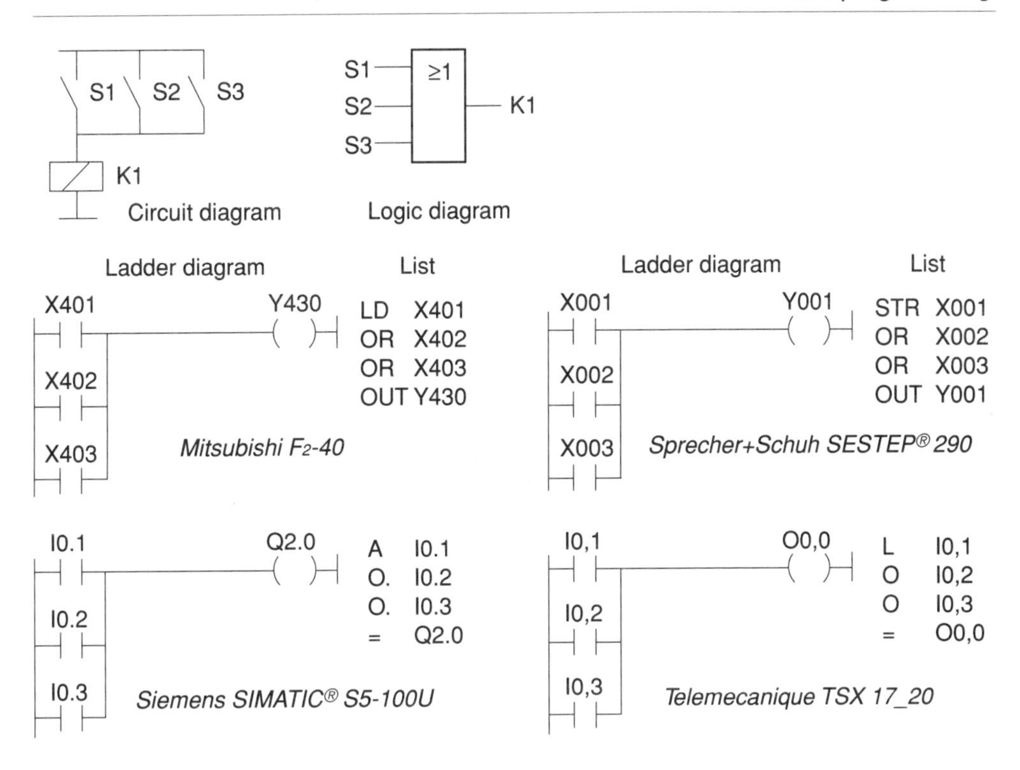

Figure 4.20. The parallel circuit (OR function).

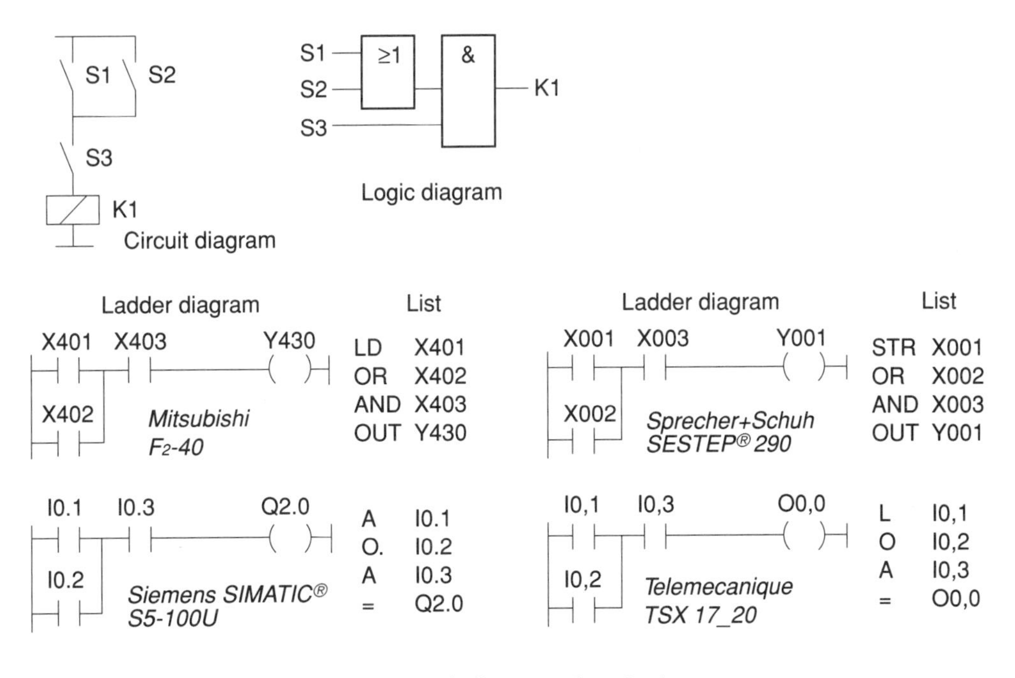

Figure 4.21. Parallel/series circuit (OR before AND function).

4.7.4 Worked example D

We require a program that turns on K1 when (S1 and S2) or S3 are closed. This corresponds to the series/parallel electrical circuit and to the logic AND before OR function, both shown in Fig. 4.22.

The conditions:

- a signal from S1 AND S2 (taken first)
- OR a signal from S3
- turns on K1.

The list programs reflect these conditions. It is important to note that, with some exceptions, the OR instruction bypasses all the preceding instructions that relate to the turning on of K1. (This is equivalent to wiring back to the upper supply line of the circuit.) The minor differences between the programming instructions are explained in examples A, B and C above.

The ladder diagrams are identical apart from the I/O names.

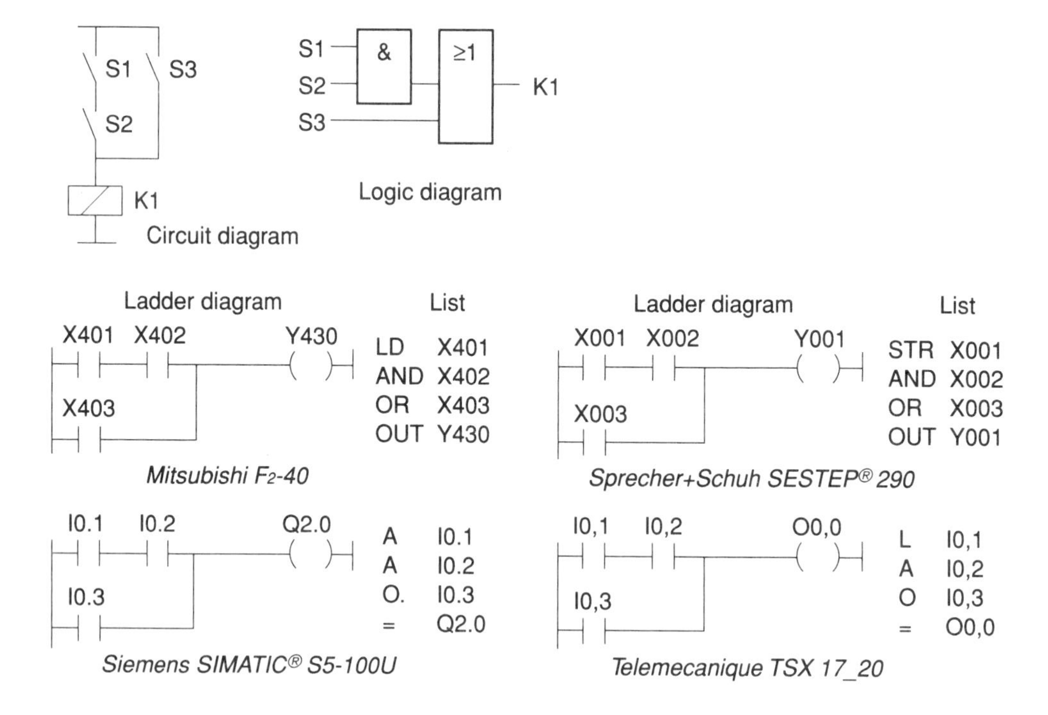

Figure 4.22. Series/parallel circuit (AND before OR function).

4.7.5 Worked example E

We require a program that turns on K1 when S1 and S2 are closed and keeps K1 turned on even when S1 is no longer closed. This is a standard scheme for motor control applications whereby pushbuttons are used for starting (S1) and stopping (S2). It corresponds to the parallel/series electrical circuit and to the logic OR before AND function, both shown in Fig. 4.23.

The conditions:

- a signal from S1 OR K1 (taken first)
- AND a signal from S2
- turns on K1.

In practice S1 has a make (normally open) contact that supplies an input signal when the button is depressed. S2 has a break (normally closed) contact that supplies an input signal at all times *except* when the button is depressed. A brief signal from S1 starts K1, which, because it is programmed as a parallel branch (OR function), can keep itself turned on as long as the signal from S2 is present. Note that it is perfectly acceptable to test the *output* in the same way as an input. Thus the ladder symbol for testing '—| |—' can be associated with outputs also.

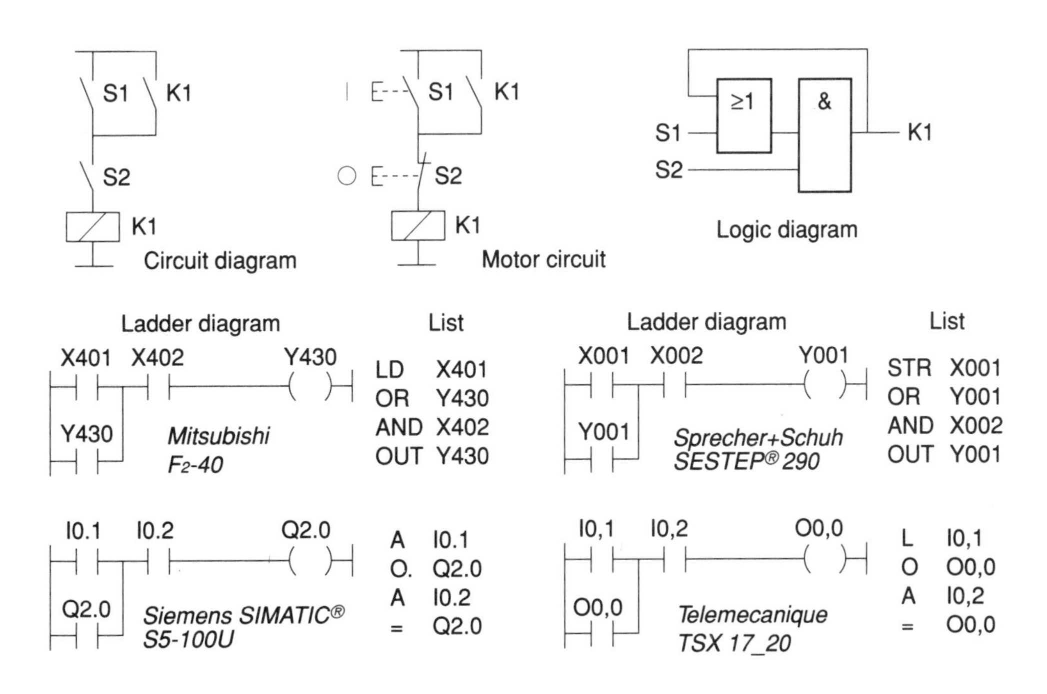

Figure 4.23. Hold-on circuit (latch function).

4.7.6 Worked example F

We require a program that turns on K1 when S1 and S2 are closed and S3 is open. This corresponds to the series electrical circuit and to the logic AND function, both shown in Fig. 4.24.

The conditions:

- a signal from S1
- AND a signal from S2
- AND NOT a signal from S3
- turns on K1.

The four list programs show the INVERSE AND instruction in different ways: Mitsubishi, **ANI**; Siemens and Telemecanique, **AN**; Sprecher+Schuh, **AND NOT**.

The ladder diagrams are identical apart from the I/O names. The symbol '—| / |—' tells the controller that there must be *no signal* from S3 to turn on K1.

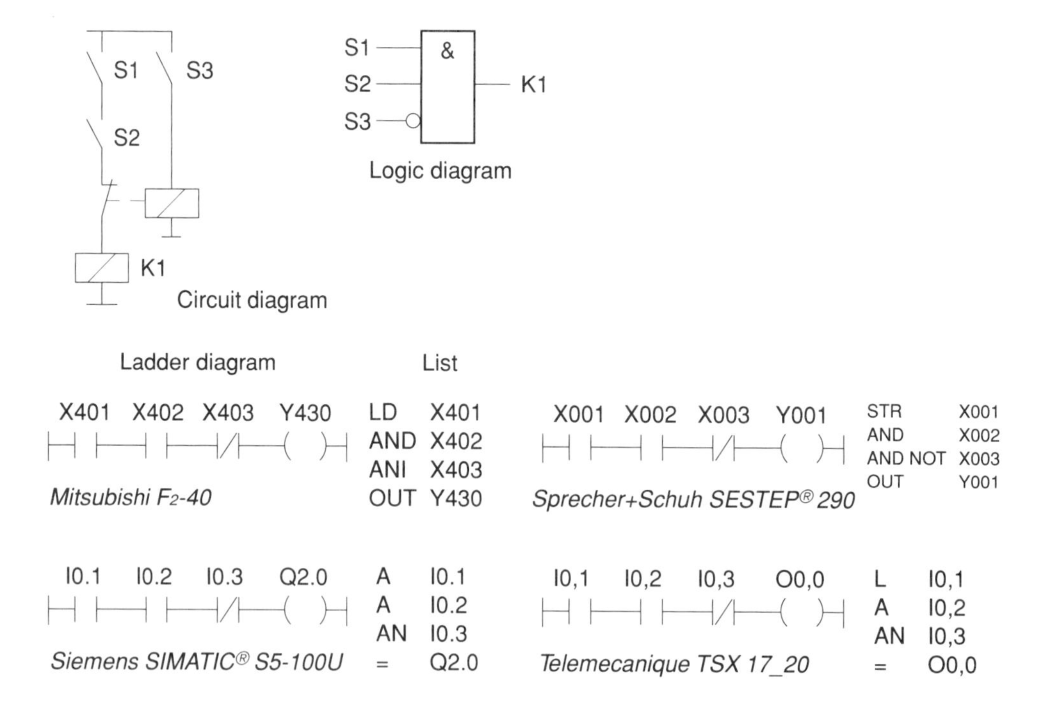

Figure 4.24. Circuit where a signal must be absent for operation (inverse function).

4.8 Guidelines for programming

1. Inputs, outputs and other devices can be tested as often as necessary. Thus we may encounter symbols such as '—| |—' repeatedly in a ladder diagram.
2. An output can be turned on only once. Specifying two or more 'turn on' commands for a given output, each with different conditions, leaves the processor in a dilemma: which set of conditions is to be used? In a ladder diagram we generally encounter the symbol '—( )—' only once in respect of each output.
3. In list programming, remember to use the specified start instruction (**LD**, **A**, **STR** or **L**) when you start entering the logic for the next output.
4. In list programming, remember to use the specified turn-on instruction (**OUT** or =) when programming an output.
5. Use only the allocated I/O names and instructions for the particular model and manufacture of controller that you are using.
6. Remember that there is *no connection* between the type of contact on an input device and the ladder symbol for testing it in the program. Both make contacts (normally open) and break contacts (normally closed) can be checked for a signal present '—| |—' or for a signal not present '—| / |—', depending on the control requirements.
7. Enter the program from a prepared draft copy, not from your head!
8. Even though programming terminals have excellent facilities for inserting, deleting and modifying programs, it is better, by using care and good planning in the first instance, to aim for the minimum of alterations.

4.9 Exercises for Chapter 4 (Solutions in Appendix 1)

4.1. Refer to Figs 4.5 and 4.6. What is the effect of closing S3?

4.2. Figure 4.20: a new switch S4 is to be included. When S4 is closed the other switches have normal control; when S4 is open the output cannot be turned on.

4.3. Figure 4.21: modify the program such that switch S3 must be open for the output to be turned on.

4.4. Figure 4.22: a new contactor K2 is to be added: the contactor is to be controlled by a switch S4 when K1 is energized.

4.5. State one advantage and one disadvantage of RAM.

4.6. Why is the 'low battery' lamp important?

4.7. A pushbutton-type control switch has a break (normally closed) contact. Does it give a signal (a) when it is depressed or (b) when it is untouched?

4.8. Are we limited in the number of times we can test an input in a program?

4.9. Why are we permitted to turn on an output only once in a program?

Chapter 5 Heating project

5.1 Project description

This project is concerned with the control of a heating system intended to supply the process and space heating needs of a building. Suitable for use with an oil- or gas-fired boiler, it is typical of the designs found in industrial, commercial and institutional buildings. The control functions involved resemble those in other applications such as refrigeration. The installation layout is shown in Fig. 5.1.

The process water is heated by circulating the hot boiler water through the heating coil of the process water tank (also called a 'calorifier'). The space heating load is divided into two zones, A and B, each having its own radiators, room thermostat and pump. When heat is required in a given zone the relevant pump runs and circulates the hot boiler water to the radiators in that zone.

The boiler is self-contained, incorporating its own control and protective functions. The only external control required is a START STOP signal to be issued by the heating control system.

5.2 Equipment

- The process is to be automated using a programmable controller.
- The boiler E1 can be started and stopped electrically.

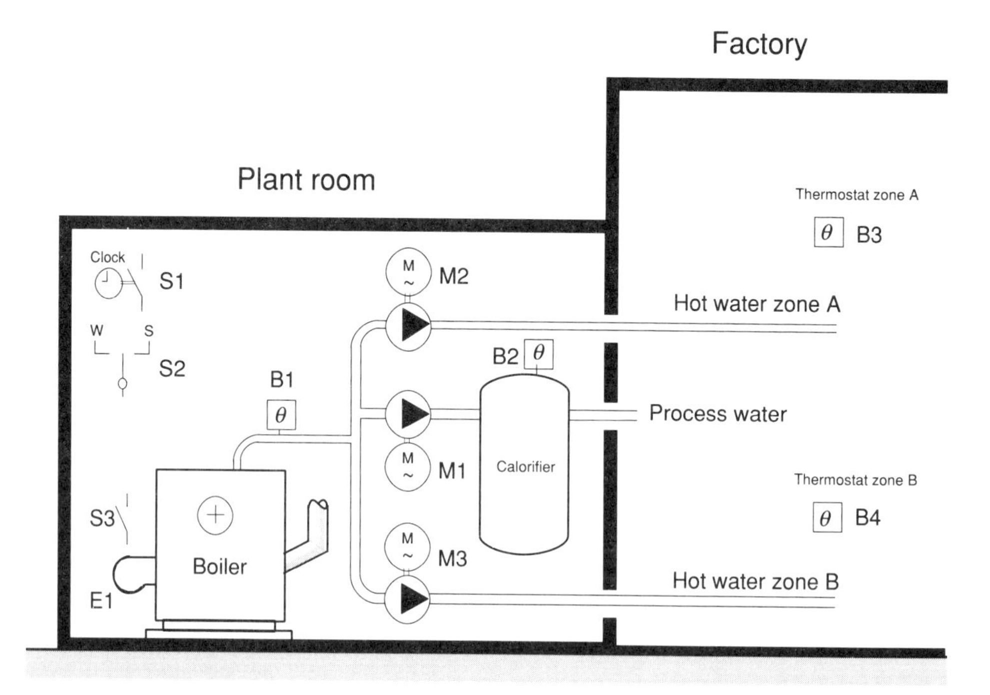

Figure 5.1. Heating project.

- Each pump motor (M1, M2, M3) has an associated contactor or relay for switching.
- A signal lamp H1 is installed to signal certain alarm conditions (see specification).
- A clock switch S1 is used for overall time control (where necessary).
- A summer/winter switch S2 allows the user to select an operating mode.
- Thermostat B1 senses boiler water temperature and B2 senses process water temperature.
- A room thermostat is fitted in each heating zone A and B (B3 and B4 respectively).
- The burner is fitted with a 'lock-out' switch which closes automatically during certain dangerous faults and is to be used for signalling.

5.3 Specification

1. The whole process is to operate only during certain hours of the day determined by the clock switch settings (or equivalent). During this time the boiler runs and heats the water to the desired temperature, regulating itself automatically.
2. When S2 is set to 'summer', only the process water is heated. When it is set to 'winter', space heating as well as process heating is provided for.
3. The process heating pump M1 runs only when the process water is below temperature (B2) and the boiler water is hot (B1).
4. The space heating pumps M2 and M3 run in response to thermostats B3 and B4 respectively, provided S2 is set to 'winter' and the boiler water is hot (B1).
5. When a burner lock-out occurs the signal lamp H1 must illuminate.

5.4 Solution

The solution can be planned in three steps, as follows.

- The main circuit, which is the same whatever controller is used. For this project we have also supplied a scaled drawing of a possible panel layout.
- The control circuit, including input and output connections, which is specific to the controller used.
- The control program, which is specific to the controller used.

5.5 Main circuit

The main circuit comprises the switchgear and protection for the pump motors and the power supply for the programmable controller. Figure 5.2 gives details.

5.6 Panel

Figure 5.3 shows one possible arrangement of the control panel for this project. It is drawn to scale (1:5) and the space allocated for the programmable controller is adequate for any of the controllers detailed in this book.

Note that the various categories of cabling (input, output and power) are allowed to be run together in the cable trunking provided they are all rated for the highest operating voltage. The programmable controller itself probably generates more heat than any other component in the panel, and accordingly is mounted in a high position.

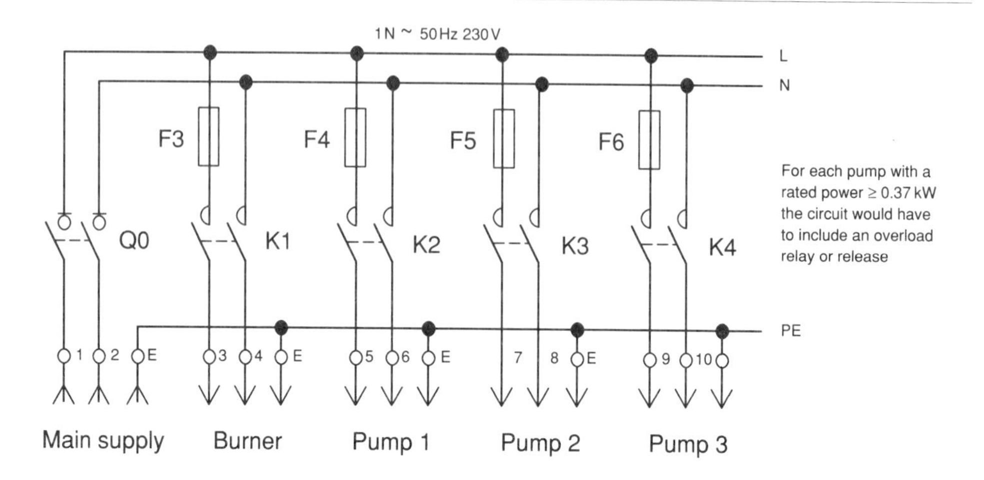

5.7 Control circuit

Tables 5.1 and 5.2 give the I/O schedules for the four controllers. The control circuits are described for each controller in Figs 5.4–5.7. In every case 24 V d.c. inputs are used; the 24 V supply is generated by the controller itself.

On the controller, one input is reserved for each thermostat and switch (a total of eight inputs). If the controller is equipped with its own real-time clock, the use of an external clock switch is unnecessary and the corresponding input is available for other uses.

Note that two inputs are required to serve S2, one for the 'summer' setting and one for the 'winter' setting. For the thermostats, our assumption in every case is that the contact is closed until the set temperature is reached; accordingly they are shown in the circuit diagram as break contacts.

One output is needed for each motor contactor, the boiler and the signal lamp (a total of five outputs). The common supply for the outputs is wired through fuse F2, which must be sized in relation to the rated current of the output relays in the programmable controller. All the outputs operate at 230 V a.c.

The connection diagram for the Mitsubishi F_2-40MR-ES (with relay outputs) is shown in Fig. 5.4. S10 and S11 have been added to STOP and RUN the program; they play no part in the operational control of the plant. On the output side, terminal COM3 is the common serving outputs Y430–Y433 and COM4 is the common serving outputs Y434–Y437. COM3 and COM4 are connected together here since all the outputs operate at the same voltage. Although we have not shown it, the connection of a snubber (*RC* network or varistor) to the coil of K1, K2, K3 and K4 is recommended to protect the output relays (refer to Fig. 3.9).

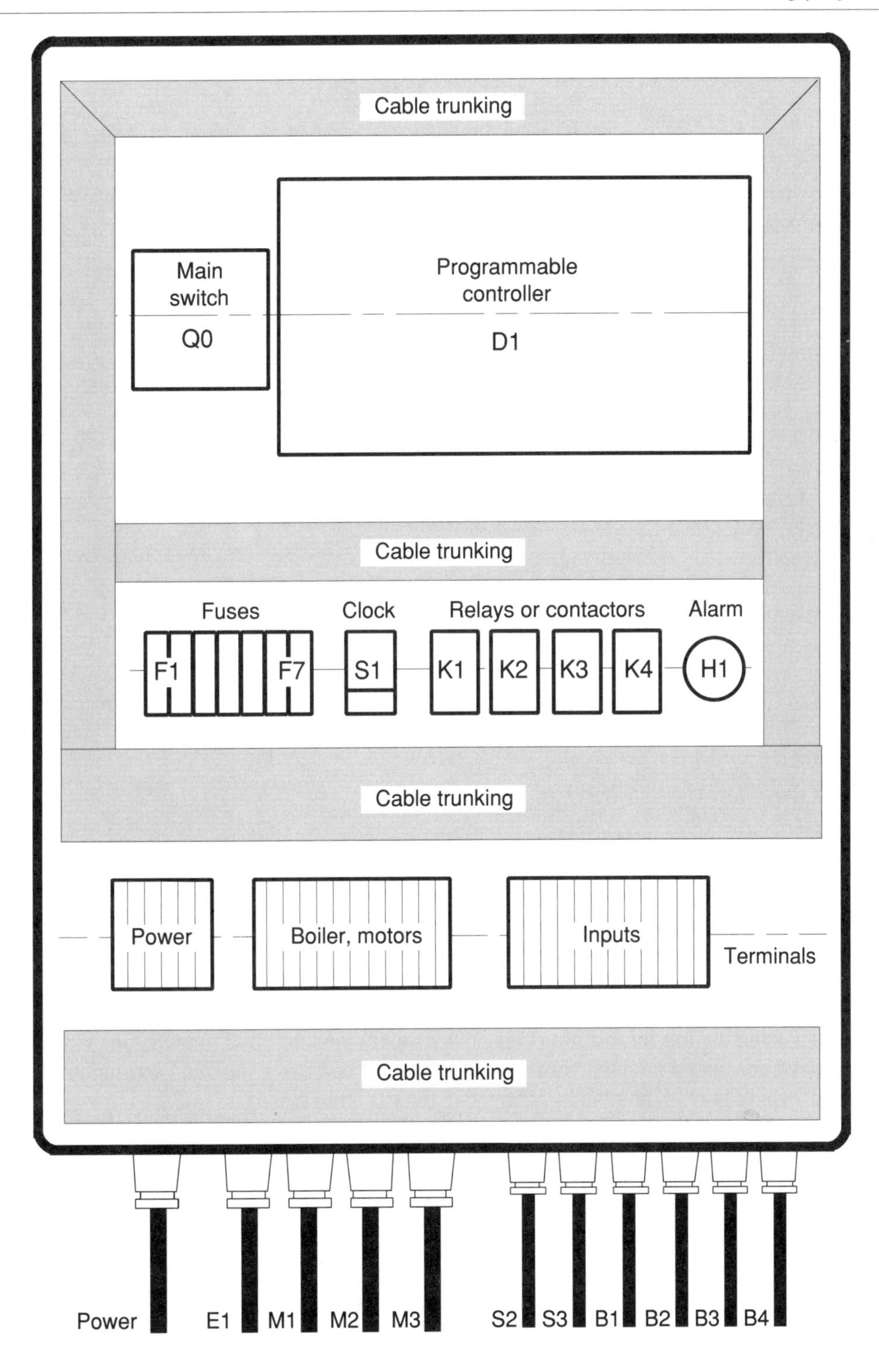

Figure 5.3. Possible panel layout.

Table 5.1. Schedule of inputs

Device	Item	Input			
		Mitsubishi	Siemens	Sprecher+Schuh	Telemecanique
Clock switch	S1	X400	I0.0	X001	I 0,0*
Summer select	S2s	X401	I0.1	X002	I 0,1
Winter select	S2w	X402	I0.2	X003	I 0,2
Burner lockout	S3	X403	I0.3	X004	I 0,3
Boiler water stat	B1	X404	I0.4	X005	I 0,4
Process water stat	B2	X405	I0.5	X006	I 0,5
Room stat A	B3	X406	I0.6	X007	I 0,6
Room stat B	B4	X407	I0.7	X008	I 0,7

* For models equipped with a real-time clock, this input need not be used.

Table 5.2. Schedule of outputs

Device	Item	Output			
		Mitsubishi	Siemens	Sprecher+Schuh	Telemecanique
Boiler	E1	Y430	Q2.0	Y001	O0,0
Pump 1 Process	M1	Y431	Q2.1	Y002	O0,1
Pump 2 Zone A	M2	Y432	Q2.2	Y003	O0,2
Pump 3 Zone B	M3	Y433	Q2.3	Y004	O0,3
Alarm lamp	H1	Y434	Q2.4	Y005	O0,4

The connection diagram for the Siemens SIMATIC® S5-100U is shown in Fig. 5.5. The minimum configuration for this project is power supply module, CPU module, one input module and one output module. Note, however, that we had two input and two output modules (as in Fig. 2.9), and this is reflected in the I/O addresses.

The 24 V d.c. supply comes from the power supply module. The positive is common and is switched by each input. Although we have not shown it, the negative must be wired to the appropriate terminal of each 24 V d.c. input module (if this is not done the input signals simply will not register).

Any of the power supply modules and any of the CPU modules is acceptable for this project (we had the 2 A power supply module and the CPU 103 module).

The output module was the relay type 6ES5 451-8MR12. The rated current of the relays in this module is 0.5 A (inductive) and the actual current for the coil of a typical small contactor is about 0.03 A, so there is no difficulty with switching capacity.

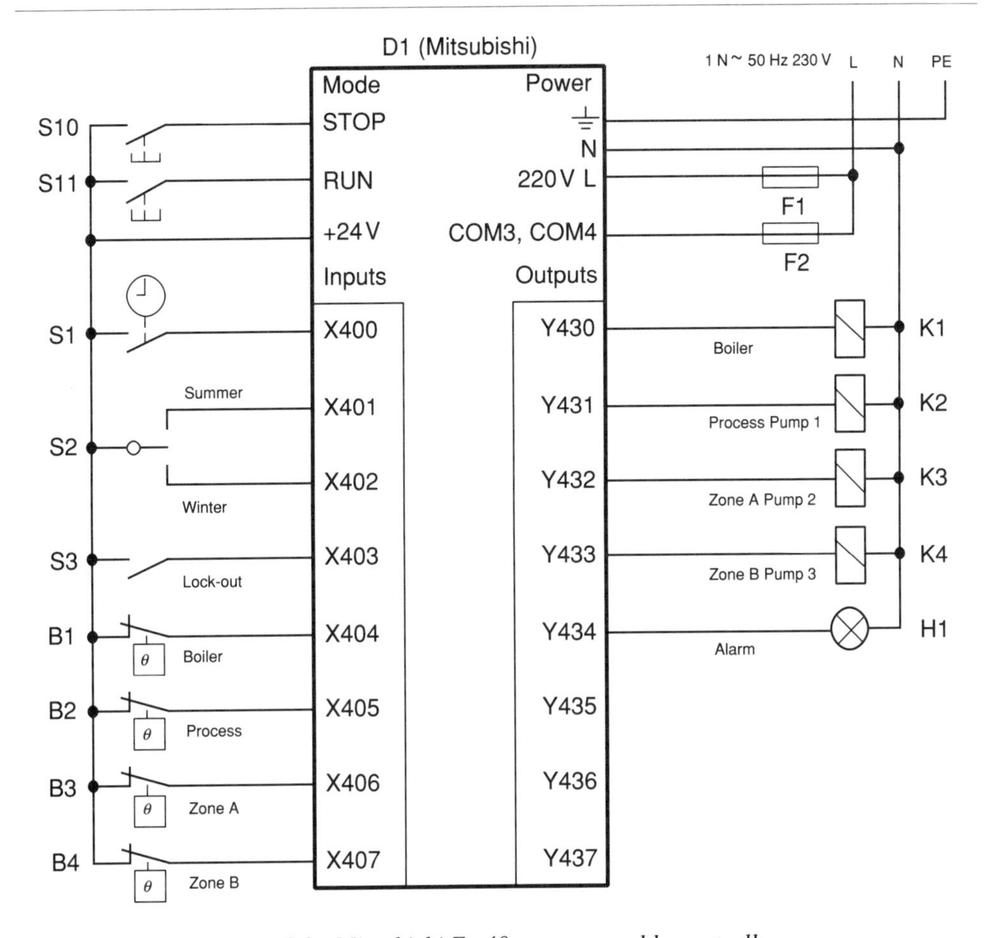

Figure 5.4. Connection of the Mitsubishi F_2-40 programmable controller.

The connection diagram for the Sprecher+Schuh SESTEP® 290 is shown in Fig. 5.6. The minimum configuration for this project is power supply module, CPU module, one input module and one output module. We had two input and two output modules (as in Fig. 2.10), but the same I/O addresses apply for all configurations.

The 24 V d.c. supply comes from the power supply module. The positive is common and is switched by each input. Although we have not shown it, the negative must be wired to the C0 and C1 terminals of each 24 V d.c. input module (if this is not done the input signals simply will not register).

The output module (relay type, ODR-21) has a more-than-adequate switching capacity for the connected loads. As with other controllers, the fitting of a snubber network to the coils of K1, K2, K3 and K4 is recommended (see Fig. 3.9).

Note that any unused slots in the rack of the controller should be blanked off with dummy modules DUM-20.

The connection diagram for the Telemecanique TSX 17_20 is shown in Fig. 5.7. The controller we used for this project had the real-time clock option fitted (micro-software

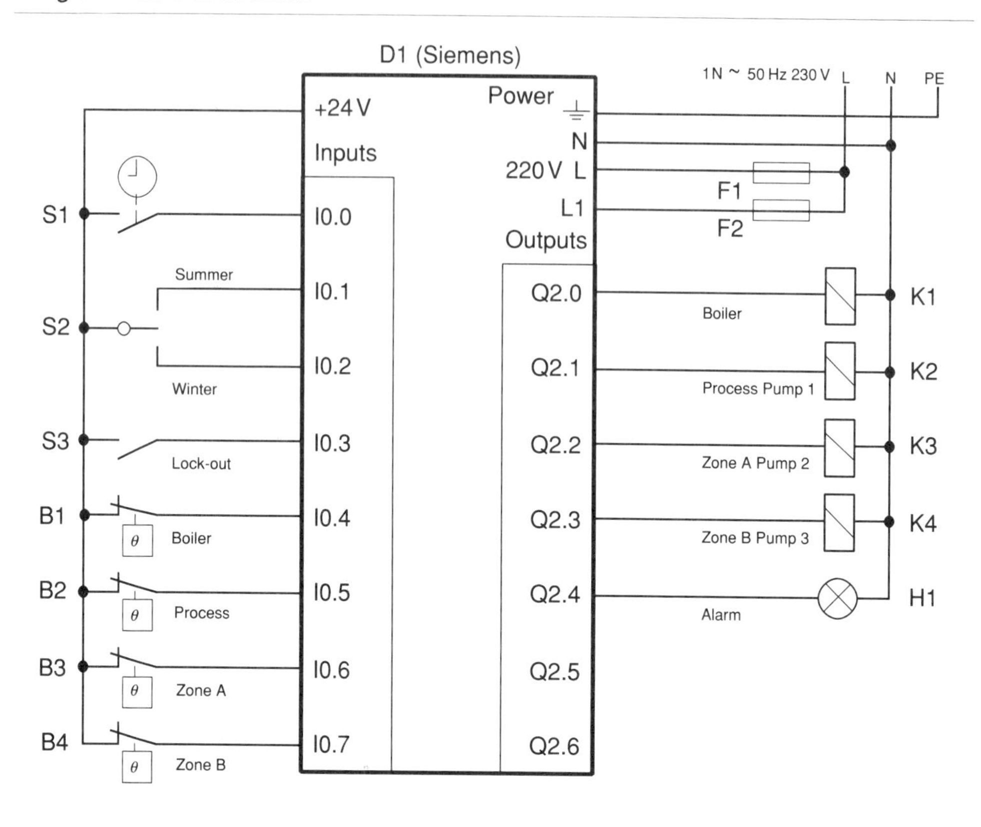

Figure 5.5. Connections of the Siemens SIMATIC® S5-100U programmable controller.

cartridge TSX P17 20FB or FD). As a result, the fitting and connection of the clock switch S1 was superfluous and input I0,0 was available for other uses.

The real-time clock enables us to program the starting and stopping times of the boiler including weekend and holiday omissions. It can also be applied to the control of any other output (the built-in calendar runs to the year 2050). Time setting is done within the program using the 'H' function block (see Fig. 5.12 and refer to Telemecanique literature for details).

Our controller (TSX 172 3428E) has relay outputs of adequate switching capacity for this application. The common terminals C0, C1, C2, C3 and C4 are connected together since all the outputs operate at the same voltage.

As with the other controllers, the use of snubber networks on the coils of K1, K2, K3 and K4 is recommended (see Fig. 3.9).

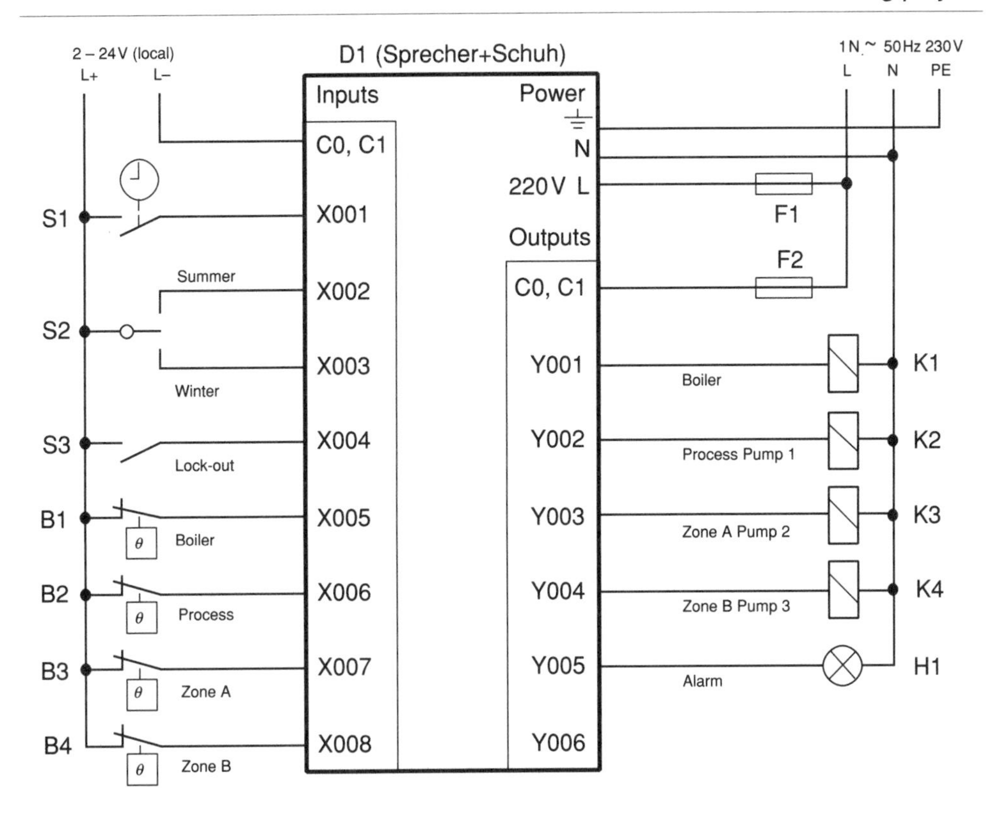

Figure 5.6. Connections of the Sprecher+Schuh SESTEP® 290 programmable controller.

5.8 Program

There are five outputs to be controlled. Thus our program will feature five sections of ladder diagram, one for each of the controlled items.

Our ladder solution may begin as a freehand sketch with only plant references (B1, H1, K2, etc.), but these can subsequently be replaced by the actual input and output numbers for the chosen controller when the solution is being formalized. We have included Fig. 5.8 as a freehand sketch for the reader to see the primitive roots of the design.

Figure 5.8(*a*) shows the boiler control. It is turned on by a signal from S2 set to winter or S2 set to summer, provided the clock switch S1 (or equivalent) allows.

Figure 5.8(*b*) shows the process pump control. It can run if there is a signal from E1 (boiler running), no signal from B1 (boiler water hot) and a signal from B2 (process water needs heat).

Figure 5.8(*c*) shows the space heating control for zone A. Pump M2 can run if there is a signal from E1 (boiler running), no signal from B1 (boiler water hot), a signal from S2 winter and a signal from B3 (zone A needs heat).

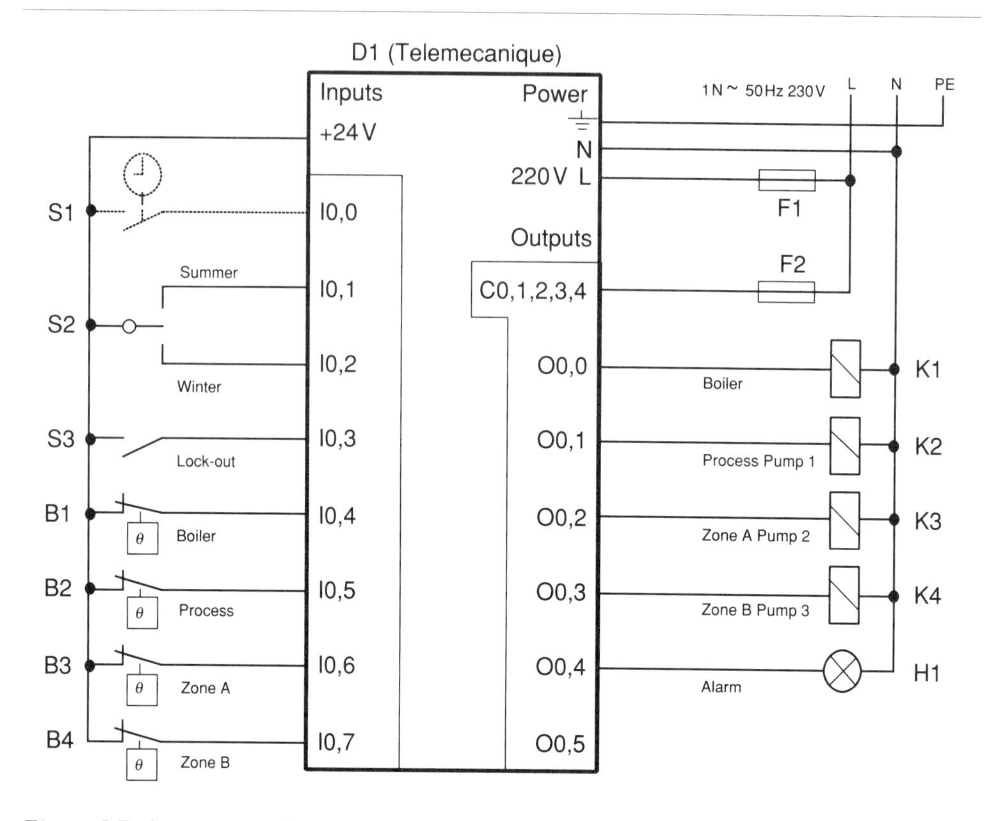

Figure 5.7. Connection of the Telemecanique TSX 17_20 programmable controller.

Figure 5.8(*d*) shows the space heating control for zone B. Pump M3 can run if there is a signal from E1 (boiler running), no signal from B1 (boiler water hot), a signal from S2 winter and a signal from B4 (zone B needs heat).

Figure 5.8(*e*) shows the alarm lamp control. It is turned on by a signal from E1 (boiler running), and a signal from S3 (lock-out activated).

Figures 5.9 to 5.12 show how the above general solution is translated into the real program for each of the four controllers. In each case the ladder diagram is broken into working sections, the size and arrangement of which is determined by the manufacturer's programming facilities.

Table 5.3 gives the corresponding list programs for the four controllers.

It is worth noting that nothing more complicated than AND and OR functions was necessary to devise the program for this project. For the newcomer, it may be encouraging to know that quite large projects are managed by breaking down the overall problem into small control tasks, each being solved in the manner indicated.

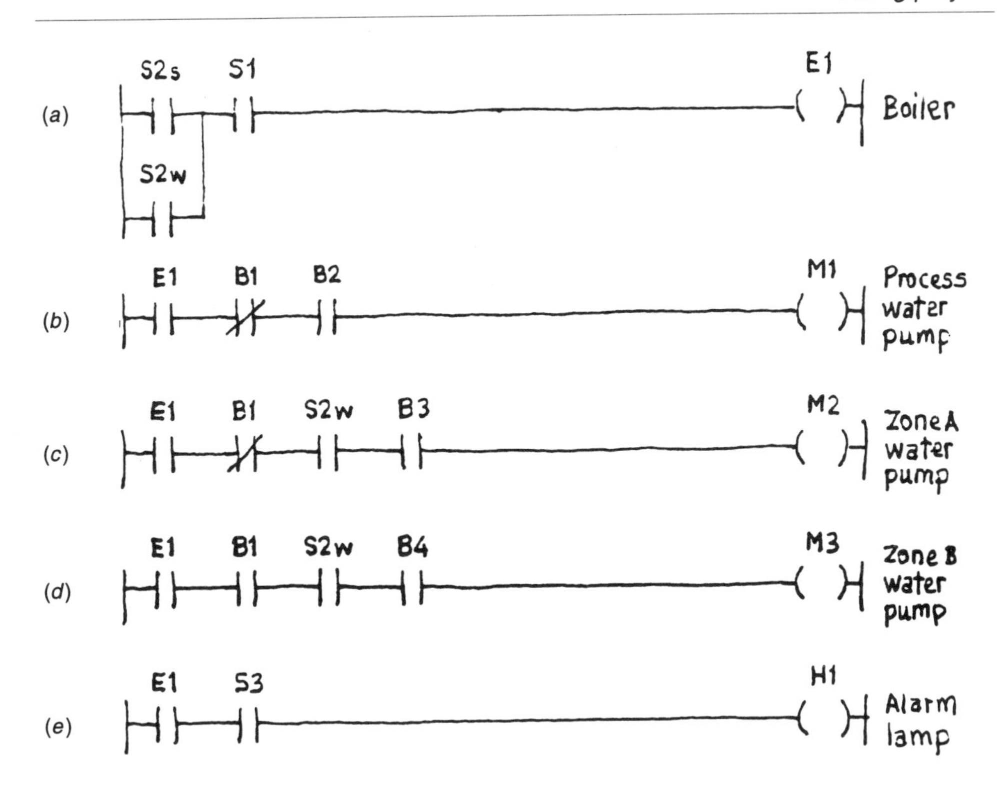

Figure 5.8. Draft solution in a generalized form.

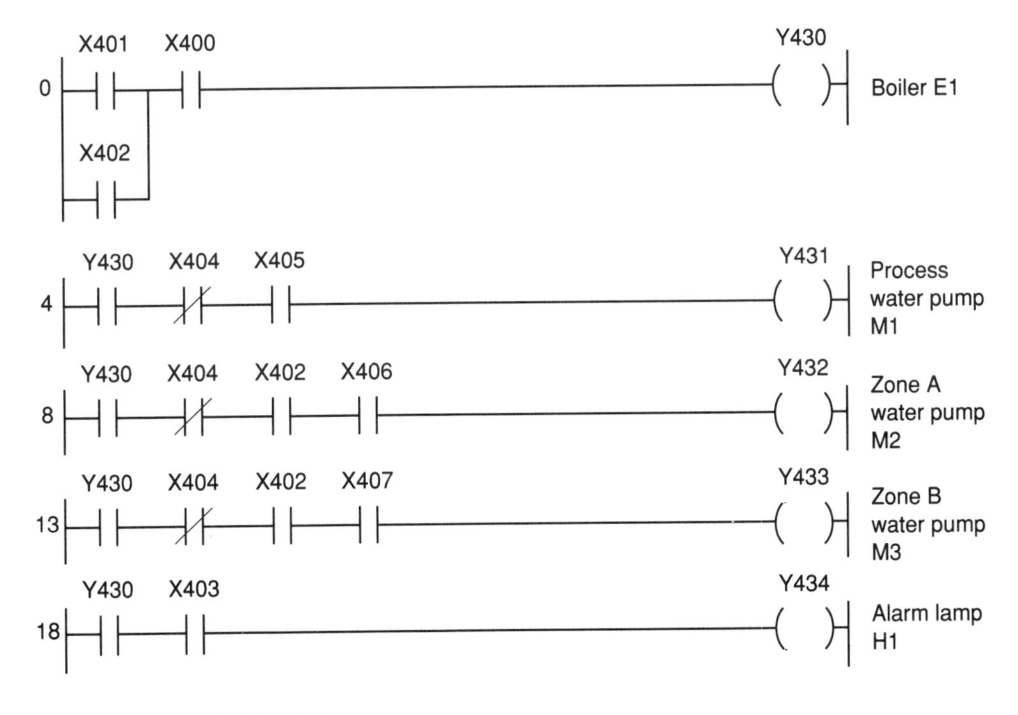

Figure 5.9. Mitsubishi solution.

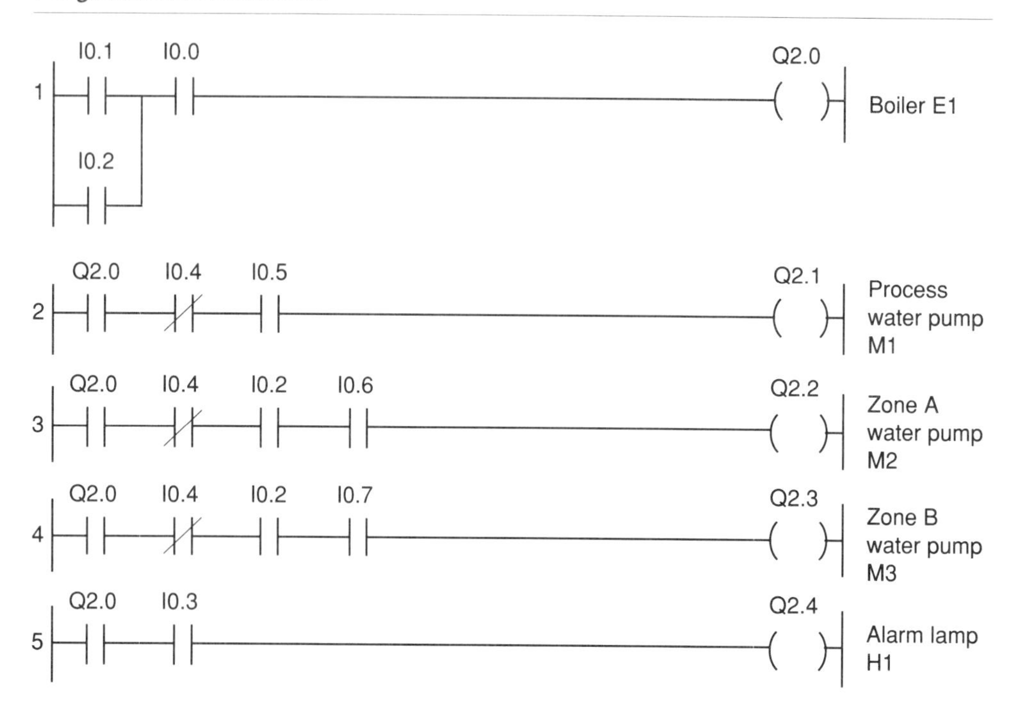

Figure 5.10. Siemens solution.

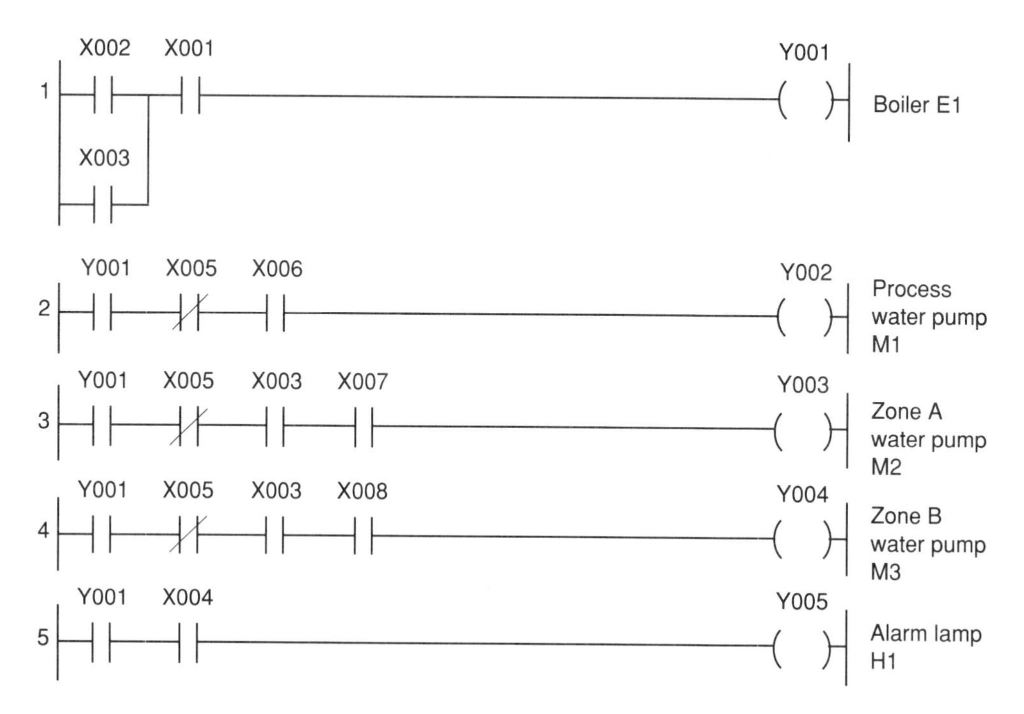

Figure 5.11. Sprecher+Schuh solution.

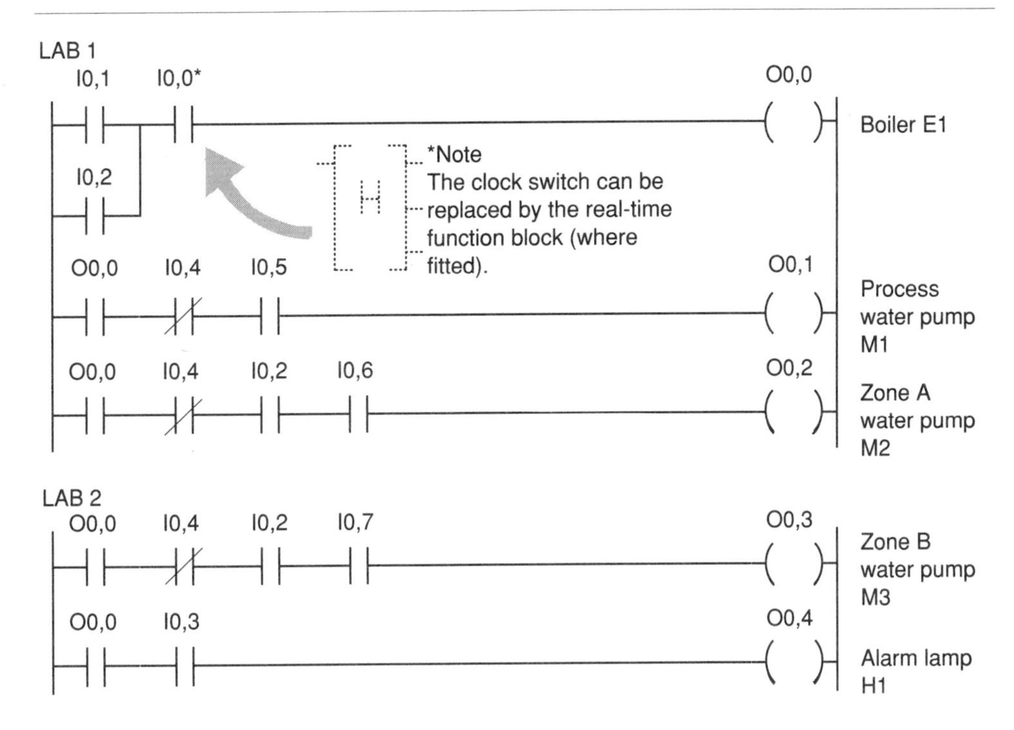

Figure 5.12. Telemecanique solution.

5.9 Implementation

- The hardware must be purchased, assembled and wired in a suitable panel. This panel must then be located in a safe position in the plant. Drawings must be available.
- The plant wiring must be carried out, the cables being terminated at the input and output terminal racks provided in the panel (refer to Fig. 5.3.)
- The software must be written to meet the specification and loaded into the programmable controller. Program documentation must be available.
- On completion of the plant wiring, every input and every output device is tested to ensure that it is wired to the correct terminal in the panel and that its signal is generated correctly. A check-list should be used for this purpose and any changes must be recorded.
- Every installation procedure recommended by the programmable controller manufacturer must be followed before powering up.
- The plant must be run on a trial basis (see the simulation procedure in Section 5.10). Any modifications must be made and recorded.
- A full production run must be made. Any necessary modifications must be made and recorded.
- The final program is stored and a backup copy made (e.g. on EPROM or diskette or cassette tape). Final documentation is made and all drawings are updated.

Table 5.3. List programs

Mitsubishi		Siemens		Sprecher+Schuh		Telemecanique	
LD	X401	A	I0.1	STR	X002	L	I0,1
OR	X402	O.	I0.2	OR	X003	O	I0,2
AND	X400	A	I0.0	AND	X001	A	I0,0
OUT	Y430	=	Q2.0	OUT	Y001	=	O0,0
LD	Y430	A	Q2.0	STR	Y001	L	O0,0
ANI	X404	AN	I0.4	AND NOT	X005	AN	I0,4
AND	X405	A	I0.5	AND	X006	A	I0,5
OUT	Y431	=	Q2.1	OUT	Y002	=	O0,1
LD	Y430	A	Q2.0	STR	Y001	L	O0,0
ANI	X404	AN	I0.4	AND NOT	X005	AN	I0,4
AND	X402	A	I0.2	AND	X003	A	I0,2
AND	X406	A	I0.6	AND	X007	A	I0,6
OUT	Y432	=	Q2.2	OUT	Y003	=	O0,2
LD	Y430	A	Q2.0	STR	Y001	L	O0,0
ANI	X404	AN	I0.4	AND NOT	X005	AN	I0,4
AND	X402	A	I0.2	AND	X003	A	I0,2
AND	X407	A	I0.7	AND	X008	A	I0,7
OUT	Y433	=	Q2.3	OUT	Y004	=	O0,3
LD	Y430	A	Q2.0	STR	Y001	L	O0,0
AND	X403	A	I0.3	AND	X004	A	I0,3
OUT	Y434	=	Q2.4	OUT	Y005	=	O0,4
END				END			

5.10 Simulation

Table 5.4 gives the steps to be followed for simulating this project. The simulation procedure described is not exhaustive, but it is systematic. It is preferable that the steps be carried out in the order indicated, but step 16 can be tried at any time and the expected result should occur.

In the interests of an impartial assessment the simulation should be undertaken by a person not directly involved in the program design, working only from the simulation list and observing the programmable controller's responses.

5.11 Exercises for Chapter 5 (Solutions in Appendix 1)

5.1. Was it absolutely necessary to use two inputs for summer/winter selection?

Table 5.4. Simulation steps for heating project

Step	Condition to be set	Remarks
1	S1, S2, S3 off. B1, B2, B3, B4 on	Manual settings, cold conditions
2	S2 set to winter. RUN the program	Program runs but no output on
3	S1 set to on (clock switch)	Boiler E1 runs
4	B1 set to off (boiler up to temp)	Pumps M1, M2 and M3 run
5	B2 set to off (calorifier water hot)	Pump M1 stops
6	B2 set to on (calorifier water cool)	Pump M1 starts again
7	B3 set to off (zone A warm)	Pump M2 stops
8	B3 set to on (zone A cool)	Pump M2 starts again
9	B4 set to off (zone B warm)	Pump M3 stops
10	B4 set to on (zone B cool)	Pump M3 starts again
11	S1 set to off (clock switch)	All outputs off
12	S2 set to summer	No effect
13	S1 set to on, B1 still off	Boiler E1 runs, Pump M1 runs
14	B2 set to off (calorifier water hot)	Pump M1 stops
15	B3 and B4 set to off (zones cool)	No effect
16	S3 set to on (at any time)	Alarm operates

5.2. What alterations would be necessary if the pumps were 0.55 kW?

5.3. How would the program be altered if the action of the thermostats were opposite to that assumed (in Section 5.7)?

5.4. For users of programmable controllers fitted with real-time clock option: remove the clock switch input and introduce the clock function block to the program. Provide the following settings: Monday–Friday, turn on 06.00–16.00.

Chapter 6 Further programming

6.1 Other elements and functions

Up to now we have focused only on the input and output elements and on the basic functions such as AND and OR. Modest though they are, they are all we need to tackle simple control problems such as those in Chapter 5.

The programmable controller can, however, deal with many more elements than inputs and outputs, and it has a variety of built-in functions that bring tremendous programming power to bear on the control problem. Examples are

- internal relays
- flip-flops
- timers
- counters
- one-shots
- sequences.

6.2 Internal relay

An internal relay is similar to an output, but there is no physical access to it. Within the program it is turned on using the symbol '—( )—' and it can be tested with the symbols '—| |—' or '—| / |—'. A programmable controller typically has hundreds of internal relays. They are used to store intermediate logic results, or for economizing on the number of steps used in the program.

The internal relay is known by the following names and numbers

Mitsubishi	'Auxiliary element'	M100, M101, M102, M103...
Siemens	'Flag'	F0.0, F0.1, F0.2, F0.3...
Sprecher+Schuh	'Coil'	C001, C002, C003, C004...
Telemecanique	'Bit'	B0, B1, B2, B3...

Figure 6.1 shows one of the many applications for an internal relay (the I/O names are not specific to any manufacturer). It refers to an automatic machine that can be started and stopped using pushbutton-type switches. The machine has several outputs which may be turned on only if the machine has been started. Rather than try to build the START–STOP control directly into the logic for every output, we let it control one internal relay instead. Later, we include this relay in the conditions for turning on the various outputs.

An internal relay may be designated *retentive* by the user. This affects only its behaviour after a power cut—a retentive relay automatically resumes the state it held prior to the loss of supply (battery backup). This is useful in applications where a cycle must be preserved in spite of power cuts.

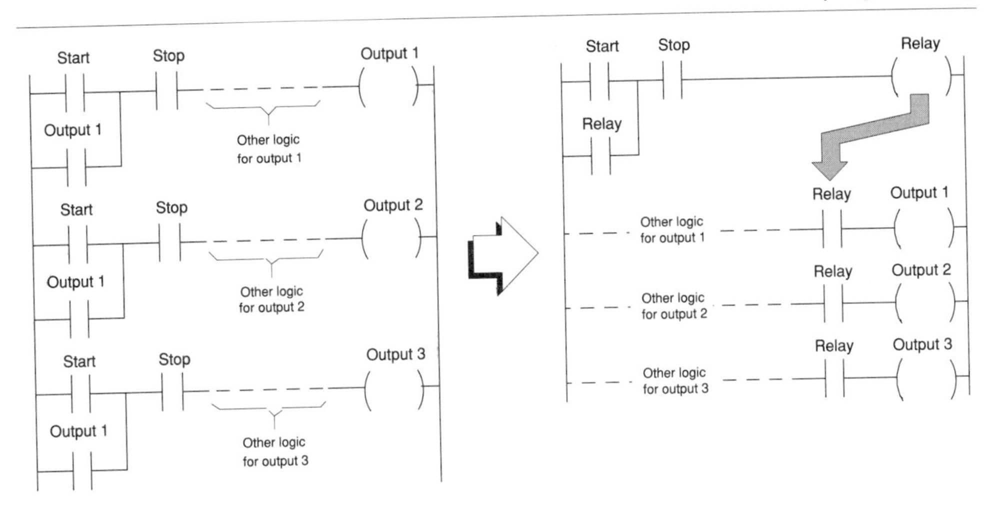

Figure 6.1. Internal relay.

6.3 Flip-flop

A flip-flop is a two-stage action or function that can be applied to an output or an internal relay. The turning ON and OFF actions work separately and both have a latching effect. The turn-on action is known as *set* and the turn-off action as *reset*. We generally use a flip-flop when we want a brief signal to produce a longer lasting result.

Figure 6.2 shows a simple implementation of a flip-flop in which one input is used to set an output and a second input is used to reset it. A signal at the first input causes the output to be turned on; the first input is no longer needed now, since the output has been set (latched on). A signal at the second input causes the output to be turned off; the second input is no longer needed now, since the output has been reset (latched off).

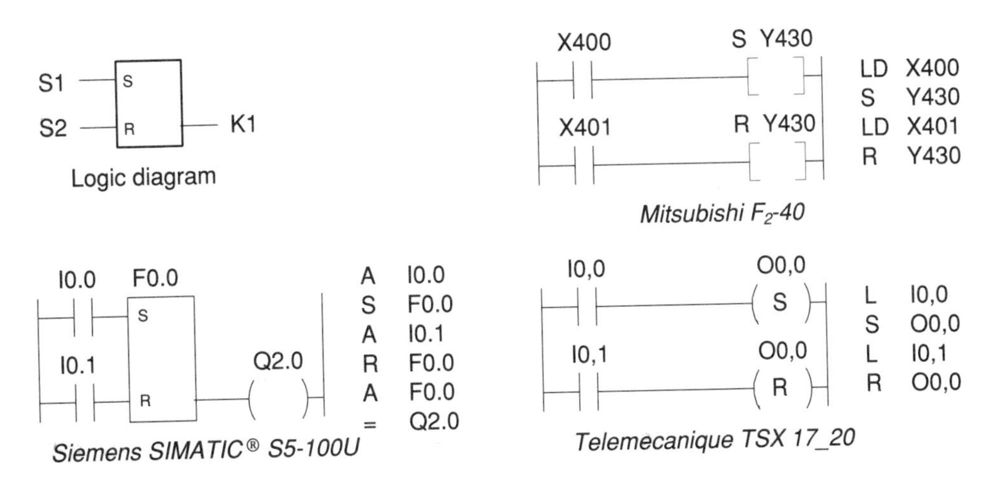

Figure 6.2. Flip-flop.

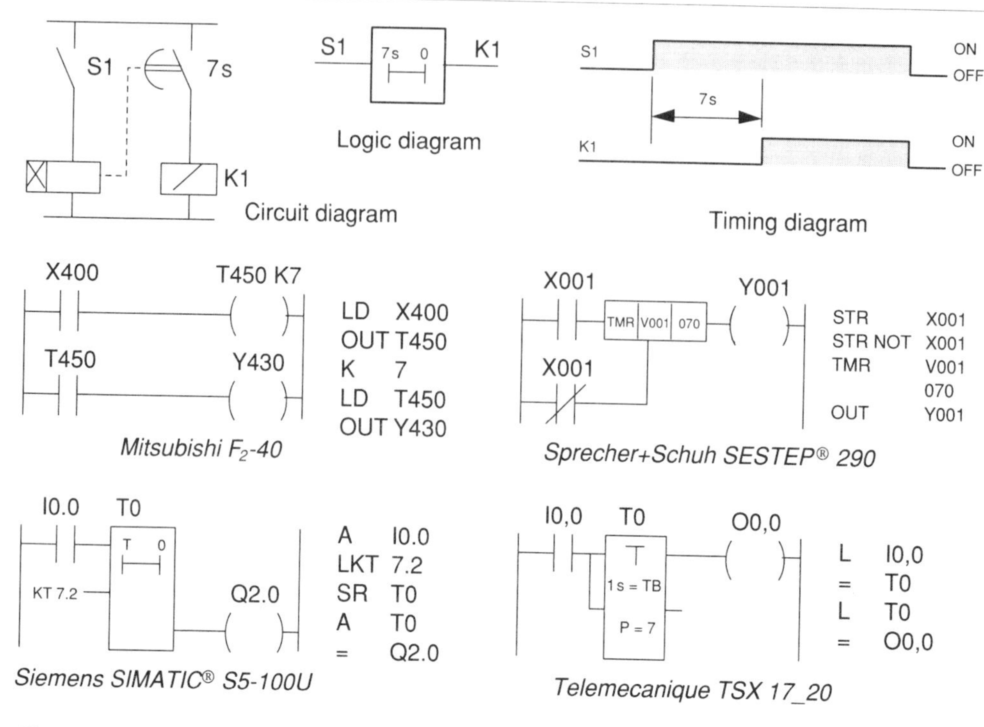

Figure 6.3. On-delay timing.

Some manufacturers use block symbols and others use modified output symbols to represent the flip-flop in the ladder diagram. The new IEC standard specifies '—(S)—' for set and '—(R)—' for reset.

6.4 Timer

Two timing functions are widely used: (a) on-delay and (b) off-delay. An on-delay timer introduces a delay between the *starting* of one item and the starting of another. An off-delay timer introduces a delay between the *stopping* of one item and the stopping of another.

Figure 6.3 refers to the on-delay timing function; the timing diagram explains its behaviour. At a given instant S1 is turned on and 7 s later K1 is turned on. When S1 is turned off, K1 is also turned off.

Figure 6.4 refers to the off-delay timing function. The timing diagram explains its behaviour: when S1 is turned on, K1 also turns on without delay. At a given instant S1 is turned off but K1 remains on for a further 12 s.

This function can be implemented with a purpose-made off-delay timer or with a suitably arranged on-delay timer (both shown). The latter is worth considering because many controllers are equipped only with on-delay timers. The technique is to allow S1 to turn on K1, which then holds itself on. When S1 is later switched off the timer is started; when the set time has expired K1 is switched off.

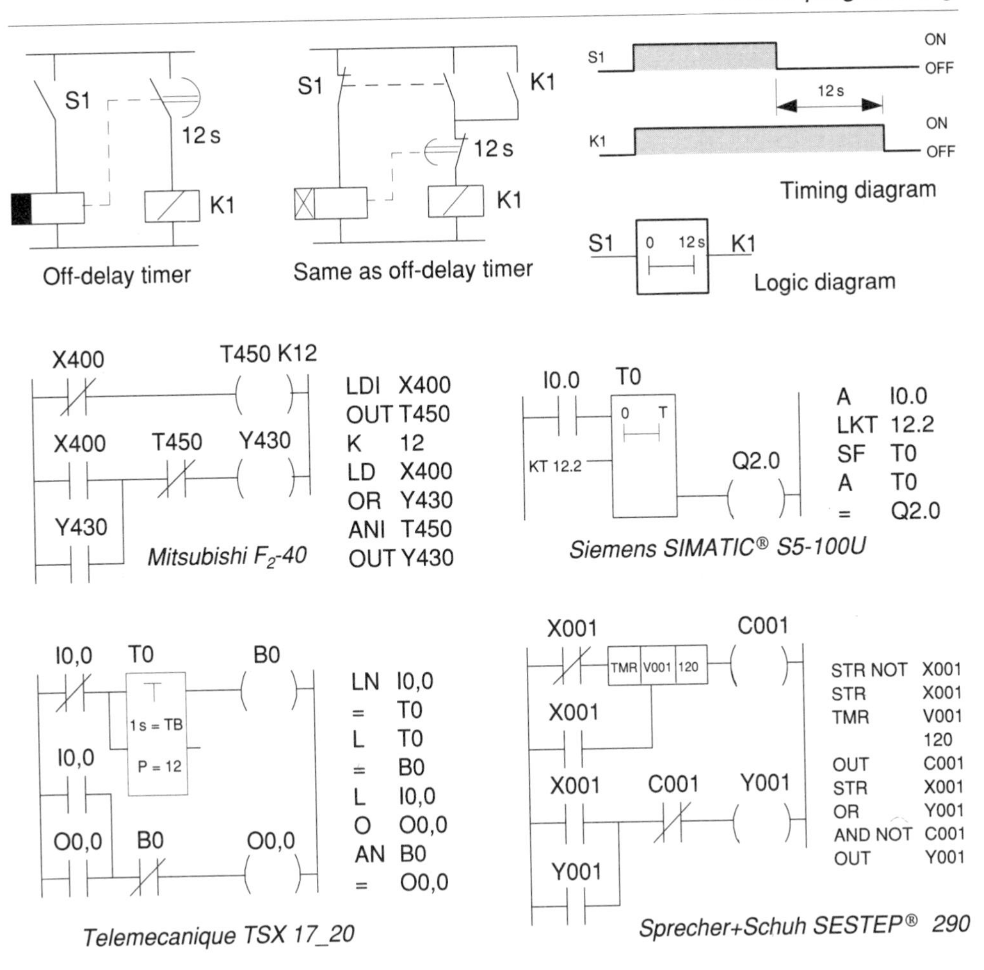

Figure 6.4. Off-delay timing.

6.5 Counter

A counter allows us to record the number of occurrences of a wide range of signals. Used in conjunction with a suitable input, it can be used to count

- the number of bottles passing along a conveyor
- the number of persons visiting a premises
- the number of revolutions of a shaft
- the number of faults in a machine.

The programming of a counter requires us to do at least three things

1. identify what signal is to be counted
2. set the target value (or pre-set value)
3. arrange the reset of the counter (its state before counting begins).

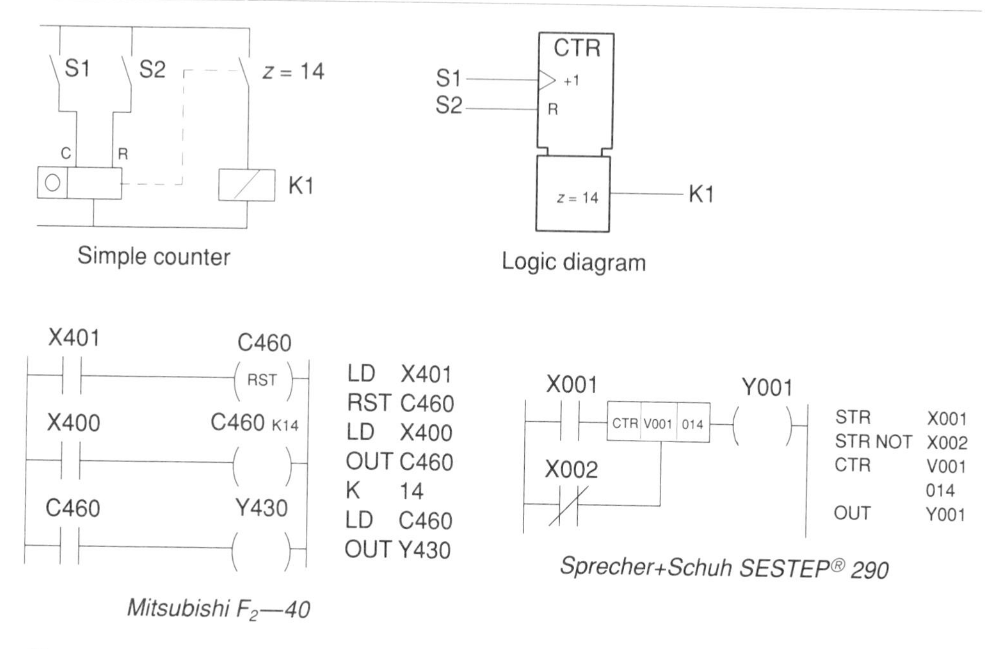

Figure 6.5. Simple counting program.

All the controllers detailed in this book are able to perform both up and down counting. However, many applications require only counting up (or down, but not both). Figure 6.5 demonstrates simple counting of this kind. In this example one input is used to count to 14, whereupon an output is turned on. A second input is used to reset the counter, whereupon the output is turned off again.

Up and down counting is illustrated in Fig. 6.6. In this case one input is used for counting up, a second is used for counting down and a third for reset. The target value is again 14.

The count-up input increments the current value of the counter; the count-down input decrements it. Although they appear similar, the two programs illustrated in Fig. 6.6 behave differently.

The Siemens counter uses a signal to load its target value—in this case we have used F0.0. The output is turned off immediately. By counting down to zero the output is turned on. The count-up input can drive the count beyond the target value (up to 999), but down-counting stops at 0. The reset input places a 0 (zero) in the counter and the output is immediately turned *on*.

In the Telemecanique counter the target value, can be loaded unconditionally. When the count equals the target value, the output is turned on. If the count-up signals continue to arrive, the output will be turned off again since the count can be driven beyond the target value (up to 9999). Down-counting does not stop at 0 but 'carries' to 9999 instead. The signal C0,D (count Done) can be used to prevent counting beyond the target value. The reset input places a 0 (zero) in the counter, which turns the output off and effectively freezes counting operations.

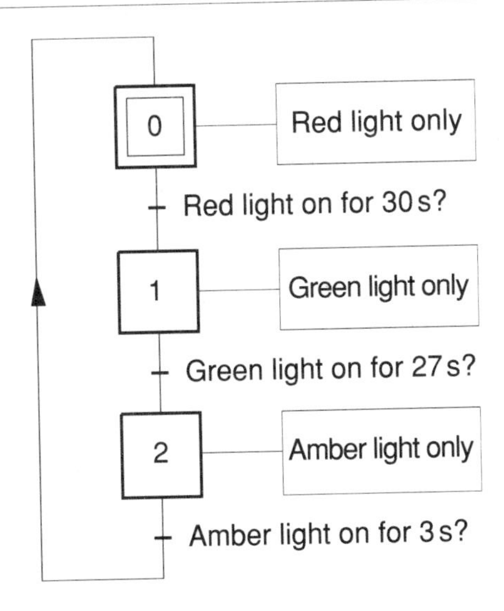

Figure 6.8. Sequential function chart.

To facilitate a trouble-free description of control sequences there exists an international graphical method called Sequential Function Chart (SFC), which is now also an approved means of organizing the program in a programmable controller. A control program based on SFC has an inherently stable structure, has a shorter scan time and is easy to troubleshoot.

The function chart for traffic light control is shown in Fig. 6.8. Each box is called a *step* and may control actions, such as turning outputs on/off. In our case, each step controls one of the traffic lights. For a light to be turned on, the relevant step must be active. Step 0 is drawn with a double box because we want it to become active immediately the program runs; such a step is called an *initial* step.

Between two steps is a *transition*—a condition which, when fulfilled, enables the next step to become active and the preceding one inactive. Inputs, outputs, timers and numerous other variables can be tested in a transition. For our traffic light control, the transitions are defined purely on a timed basis. Note that a transition is effective *only* when the preceding step is active.

Figure 6.9 shows schematically the execution of the traffic light sequence.

Figure 6.9(*a*).	Step 0 is active on power-up, the red light is on, all others are off.
Figure 6.9(*b*).	The red light has been on for 30 s, so the transition is cleared and step 1 becomes active, step 0 inactive. The green light is on, all others are off.

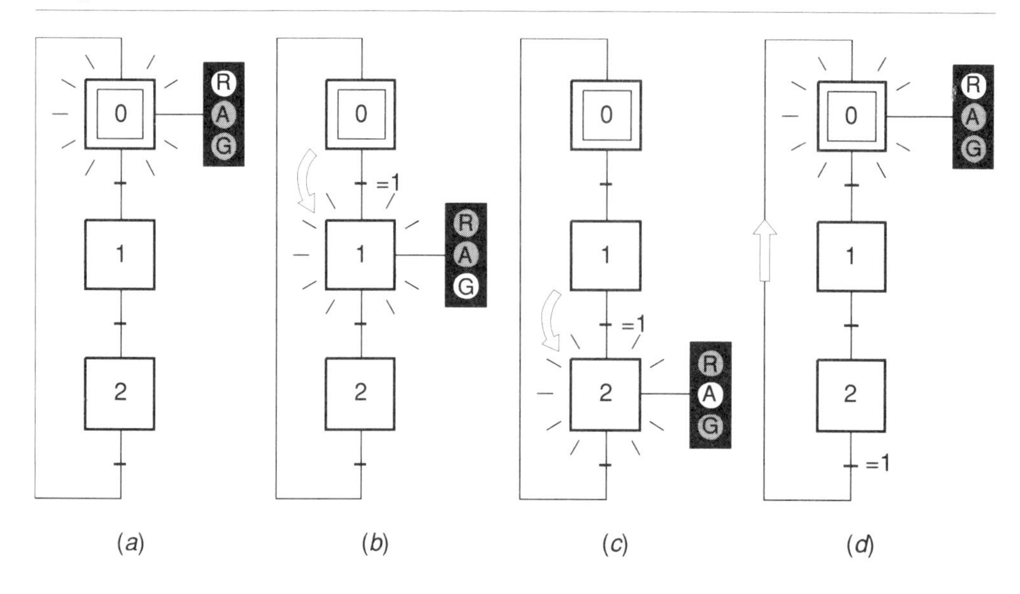

Figure 6.9. Visualizing the traffic light cycle.

Figure 6.9(*c*). The green light has been on for 27 s, so the transition is cleared and step 2 becomes active, step 1 inactive. The amber light is on, all others are off.

Figure 6.9(*d*). The amber light has been on for 3 s, so the transition is cleared and step 0 becomes active, step 2 inactive. The red light is on, all others are off and the cycle repeats.

The Telemecanique implementation of sequential function chart is known as GRAFCET. Its application to the traffic light problem is illustrated in Fig. 6.10. The chart itself is created in accordance with the principles outlined above; the link between the last and first steps is not 'drawn' directly but is implied by connectors.

The actions to be carried out at each step are programmed in ladder form after 'zooming in' to the step. The transition conditions are programmed in much the same way.

Mitsubishi's implementation of function chart is called 'Stepladder' and takes the form of a modified ladder diagram which is easy to learn. Siemens' implementation is known as 'Graph 5' and is supplied as a separate function block which can be called from within the main program.

To cater for more complex control requirements, two branching techniques are envisaged: selective and parallel, both illustrated in Fig. 6.11. Selective branching allows a *choice* of sequences; there are three in our example:

- steps 24, 25, 26, 27 and back to 24
- steps 24, 28, 29, 27 and back to 24
- steps 24, 29, 27 and back to 24.

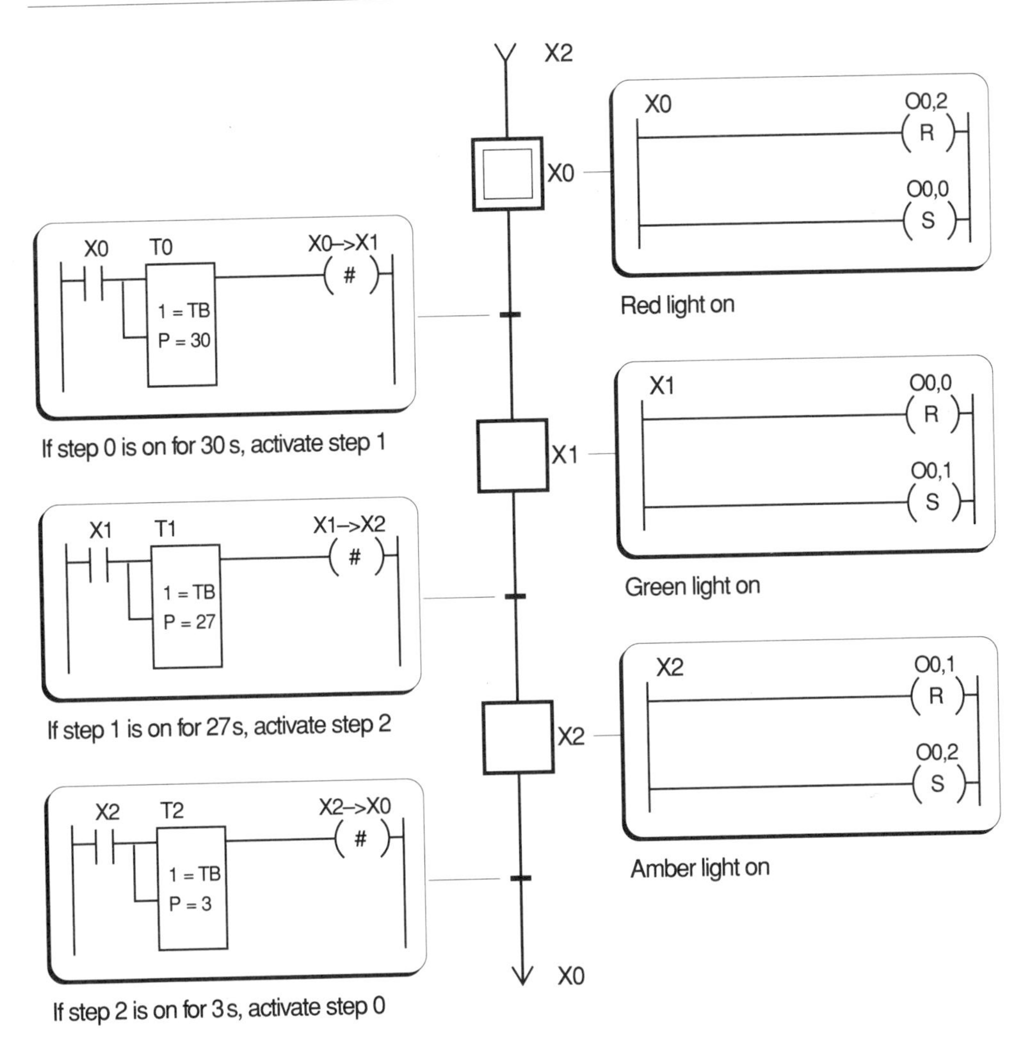

Note 1: The detail at each step or transition is obtained by using the zoom function.

Note 2: Timers T0...T2 are not essential in this application because each step has its own in-built timer for this precise task. However, we are content to use the normal timer here since it has been fully explained and is familiar to the reader.

Figure 6.10. Programming the chart, steps and transitions on Telemecanique TSX 17_20.

The choice of sequence is determined by the transitions after step 24, therefore the conditions defining these transitions have to be mutually exclusive. Selective branching allows us to conditionally skip some steps; in our example transition ‘h’ effectively causes step 28 to be skipped. Conversely, we can use it to perform conditional looping, i.e. the repeated processing of one or more steps until certain conditions are met.

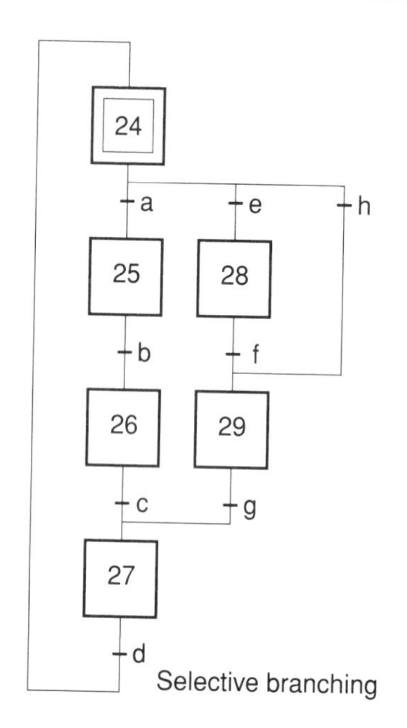

If 24 is active and 'a' is true, 25 is activated.
If 24 is active and 'e' is true, 28 is activated.
If 24 is active and 'h' is true, 29 is activated.
Conditions 'a', 'e' and 'h' must be exclusive.

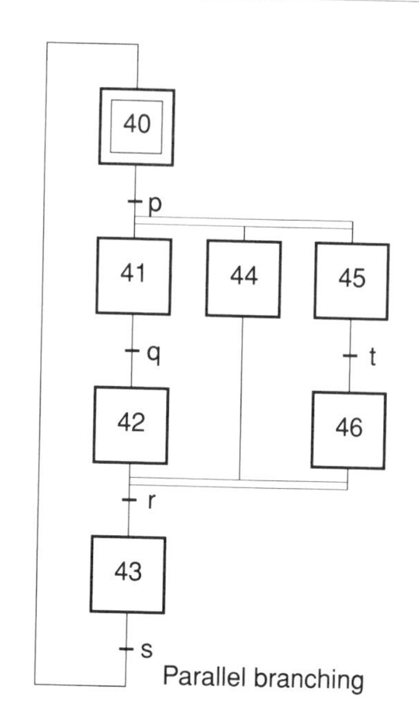

If 40 is active and 'p' is true, 41, 44 and 45 are activated together. When 42, 44 and 46 are active and 'r' is true, 43 is activated.

Figure 6.11. Selective and parallel branching.

Parallel branching allows for two or more sequences to proceed simultaneously. In our example the main sequence diverges into three sub-sequences after step 40. They are

- steps 41, 42
- step 44
- steps 45, 46.

These sub-sequences proceed independently until a convergence occurs at step 43. Note that two or more steps can be active at the same time.

6.8 Exercises for Chapter 6 (Solutions in Appendix 1)

6.1. What is a retentive relay and why is it useful?

6.2. What does 'latching action' mean?

6.3. Devise a program for flashing an output continuously. The ON time is 2 s and the OFF time 1 s.

6.4. State two advantages of sequential function chart programming.

6.5. Draw the function chart for the control of a motor having forward and reverse rotation. Pushbuttons are used to select stop, forward or reverse.

6.6. Draw a sequential function chart for traffic lights with the sequence red, red and amber, green, amber.

Chapter 7 Mill project

7.1 Project description

The process involves the transfer of animal feed from a storage silo to two production silos in which specific volumes are to be maintained. The installation layout is shown in Fig. 7.1.

After emerging through a rotary discharge valve at the base of the storage silo, the product is moved by screw conveyor to the top of the production silos where a diverter chute determines which silo is filled first. A different stage of the process determines its release from these silos.

7.2 Equipment

- The process is to be automated using a programmable controller.
- Each motor has an associated contactor for switching and a motor circuit breaker for protection (auxiliary contacts fitted).
- A hooter is installed to signal certain alarm conditions (see specification).
- A pneumatic cylinder (with associated 5/2 valve) controls the chute position.
- A speed detector is fitted to the conveyor gearbox to sense positive rotation.
- One start switch and one stop switch are used for overall control.
- Load cells and an associated amplifier signal the weight of product in storage.
- Low- and high-level sensors are fitted to each of the production silos.

7.3 Specification

1. The process is started by pressing the start switch (S1). It can be stopped at any time by pressing the stop switch (S2).
2. There is adequate product for one batch if the load cell amplifier (A1) gives a 'weight OK' signal. If insufficient product is available at the moment of starting, the hooter must sound.
3. Filling may begin once START has been called, the production silos are empty (B1 and B3), product is available (A1) and the diverter chute is moved to silo 1 (by turning on Y1 for 1 s).
4. The screw conveyor now runs (K2) and establishes a positive running signal on the speed detector (B5).
5. Once this signal is received, the rotary motor valve is activated (K1) and the product is released.
6. When silo 1 is full, the diverter is moved to silo 2 (by a 1 s signal to Y2).
7. When silo 2 is full, the conveyor and rotary valve motors both stop.
8. If the conveyor stops for any reason, the rotary valve motor must also stop.
9. An overload fault on either motor (Q1 or Q2) during production must cause the hooter (H1) to sound.
10. Automatic refilling occurs when both production silos become empty once more.

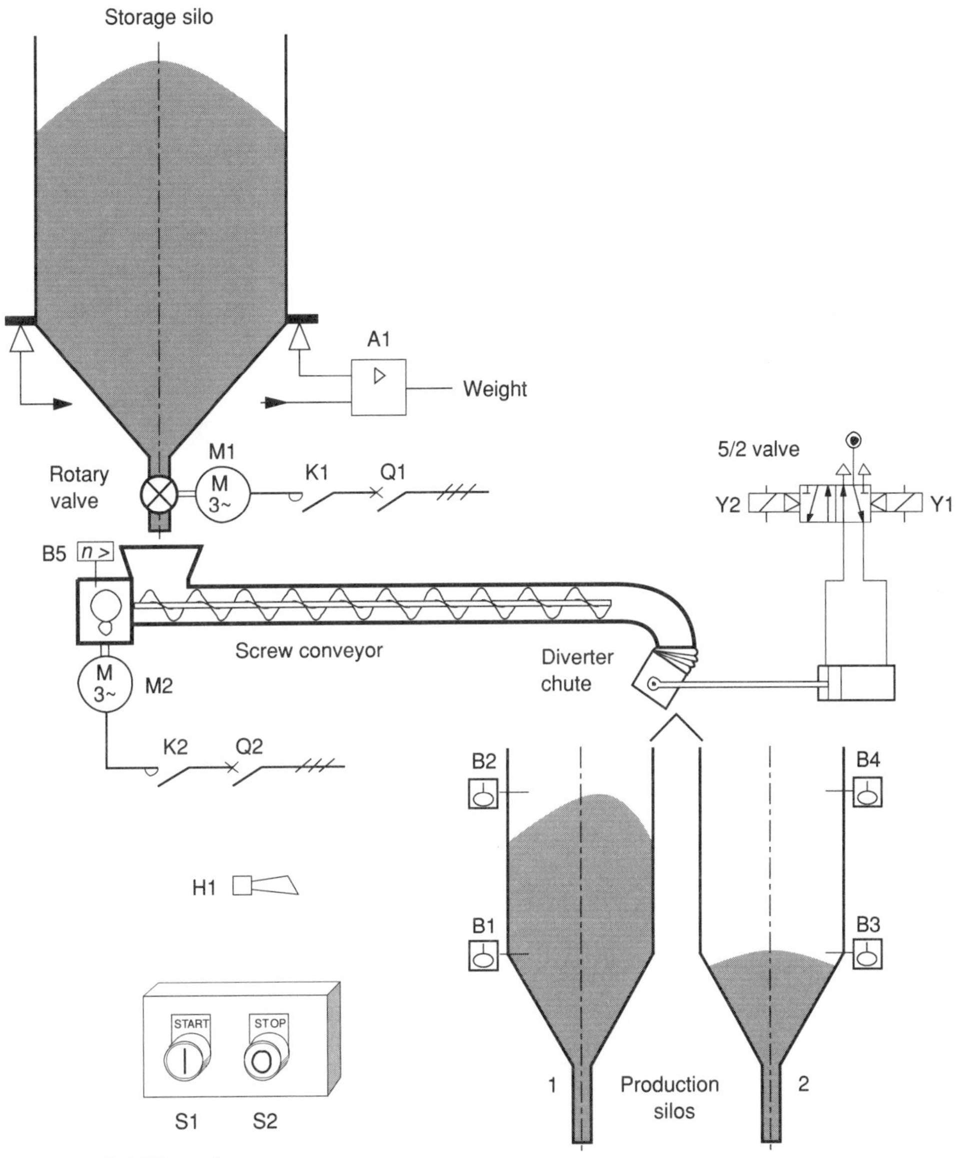

Figure 7.1. Mill project.

7.4 Solution

The solution can be planned in three steps:

- the main circuit, which is the same whatever controller is used
- the control circuit, including I/O connections, which is specific to the programmable controller used
- the control program, which is specific to the controller used.

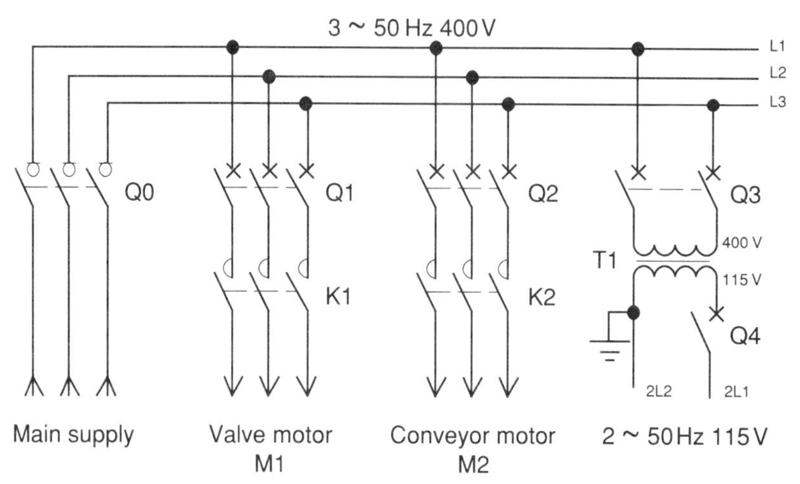

Figure 7.2. Main circuit.

7.5 Main circuit

The main circuit comprises the switchgear and protection for the motors, valves and hooter and the necessary power supplies. Figure 7.2 gives details. Control circuit transformer T1 is not essential, but by isolating the control circuit it gives a measure of protection from earth faults in the mains wiring.

7.6 Control circuit

Tables 7.1 and 7.2 indicate the inputs and outputs being used for this project. On the controller, one input is reserved for each sensor or switch (a total of 10 inputs). Auxiliary contacts belonging to Q1 and Q2 are wired as inputs to the controller to ensure that the relevant output will be turned on only when the corresponding circuit breaker is closed. The behaviour of the level sensors with the rise and fall of product must be established—for this project our assumption is that the controller receives a signal when the sensor is covered. The load cell amplifier gives a signal when the weight of the product in the storage silo is greater than a set minimum.

One output is needed for each motor contactor, the hooter and the valves (a total of five outputs). The outputs serving the motors are wired through auxiliary contacts of the relevant motor circuit breakers to assure positive safety. All the coils are rated for the chosen control voltage—in this case 115 V a.c. Miniature circuit breaker Q4 (which may be replaced by a fuse) gives protection against overcurrent in the output circuit. It must be selected with due regard for the rated current of the controller's output relays or triacs.

The connection diagrams in Figs 7.3 and 7.4 are developed from the I/O tables. In every case 24 V d.c. inputs are used; the 24 V supply is generated by the programmable controller's own power supply module.

Table 7.1. Schedule of inputs

Device	Item	Input			
		Mitsubishi	Siemens	Sprecher+Schuh	Telemecanique
Circuit breaker	Q1	X400	I0.0	X001	I0,0
Circuit breaker	Q2	X401	I0.1	X002	I0,1
Weight OK	A1	X402	I0.2	X003	I0,2
Silo 1 low-level	B1	X403	I0.3	X004	I0,3
Silo 1 high-level	B2	X404	I0.4	X005	I0,4
Silo 2 low-level	B3	X405	I0.5	X006	I0,5
Silo 2 high-level	B4	X406	I0.6	X007	I0,6
Speed detector	B5	X407	I0.7	X008	I0,7
Start switch	S1	X410	I1.0	X009	I0,8
Stop switch	S2	X411	I1.1	X010	I0,9

Table 7.2. Schedule of outputs

Device	Item	Output			
		Mitsubishi	Siemens	Sprecher+Schuh	Telemecanique
Alarm hooter	H1	Y430	Q2.0	Y001	O0,0
Rotary valve	K1	Y431	Q2.1	Y002	O0,1
Conveyor	K2	Y432	Q2.2	Y003	O0,2
Solenoid valve 1	Y1	Y433	Q2.3	Y004	O0,3
Solenoid valve 2	Y2	Y434	Q2.4	Y005	O0,4

In the Mitsubishi connection diagram (Fig. 7.3), pushbuttons S3 and S4 have been added to enable us to STOP and RUN the program. A spring-return selector switch or keyswitch could have been used instead.

Since the power supply is 115 V a.c. the user must take care to connect the mains to the correct power terminals on the controller!

Note that the operational stop switch S2 has a break (normally closed) contact which gives an input signal at all times except when depressed. Because of the controller's octal addressing, the next available input after X407 is X410.

The outputs are all inductive loads and should be fitted with snubber networks to limit the damage to the output relays in the controller. In respect of overcurrent protection for the

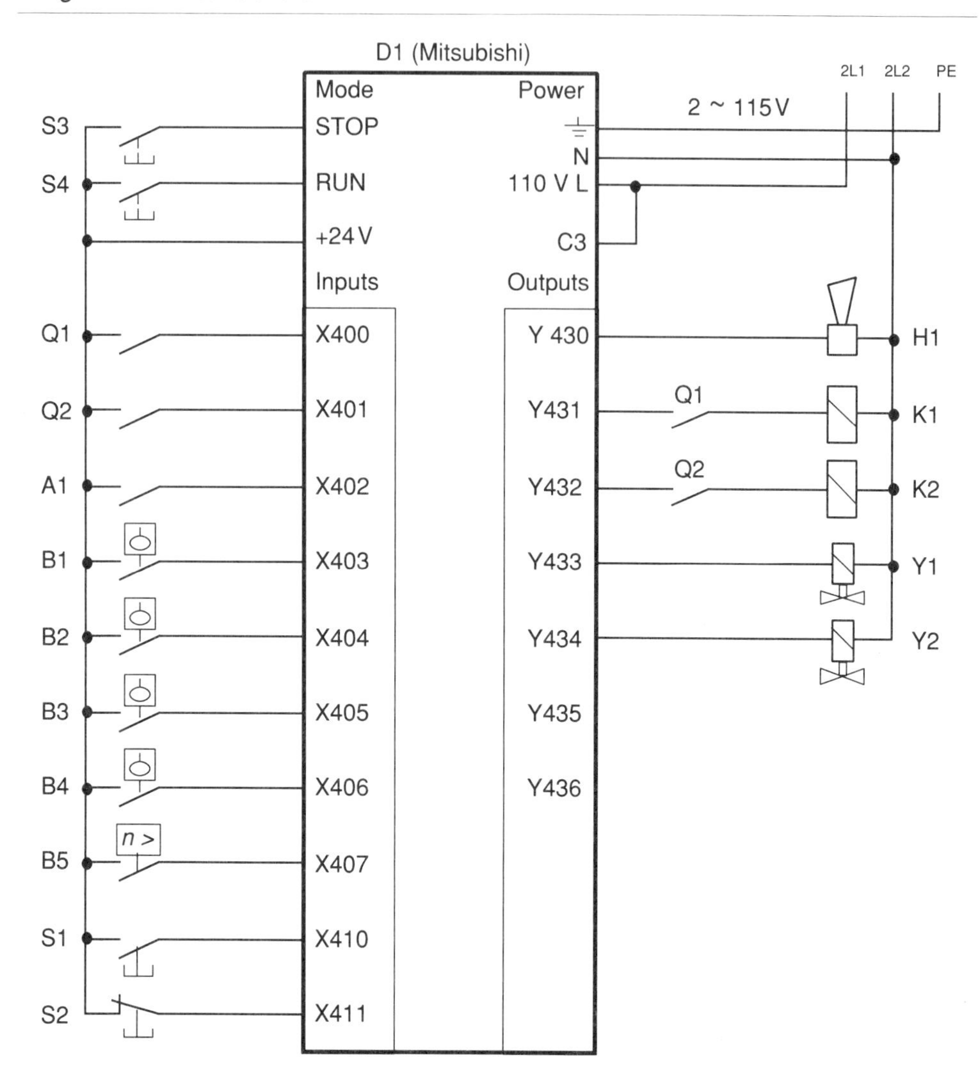

Figure 7.3. Connections for the Mitsubishi controller.

output relays, Mitsubishi recommends a fuse/miniature circuit-breaker of rated current ≤ 10 A. This limits the rated current of circuit breaker Q4 (in the circuit diagram, Fig. 7.2) to $I_n \leq 10$ A.

In the Siemens connection diagram (Fig. 7.4) we have assumed the same configuration as that adopted for all the projects and illustrated in Fig. 2.9, i.e. power supply module, CPU module, two input modules and two output modules.

The 115 V a.c. supply poses no problems for the controller, since the power supply module can automatically adapt itself to 115 V or 230 V a.c.

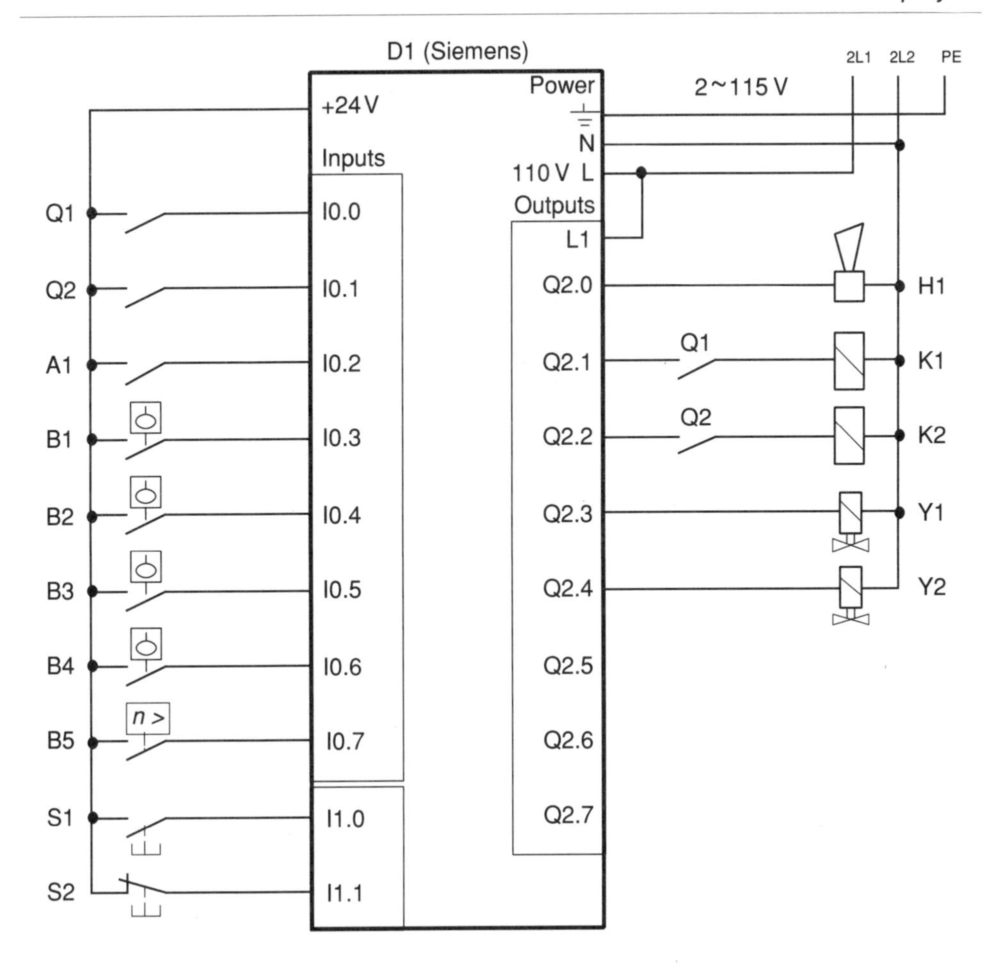

Figure 7.4. Connections for the Siemens controller.

Note that the operational stop switch S2 has a break (normally closed) contact which gives an input signal at all times except when depressed. Both S1 and S2 are connected to module 1, hence their addresses, **I1**.0 and **I1**.1 respectively.

The outputs are all inductive loads and should be fitted with snubber networks to limit the damage to the controller's output relays.

The question of overcurrent protection of the output circuits is not clearcut in this case, because Siemens makes no recommendation regarding the rated current of the fuse or circuit breaker. As a general rule, it is better to choose as low a rating as possible consistent with the normal circuit current. Each output relay is rated 0.5 A (inductive), and since there are eight outputs in the module this would seem to suggest a fuse or circuit breaker with a minimum rating 4 A.

7.7 Program

There are five outputs to be controlled. However, we also have to manage the overall control using (spring-return) start and stop switches, and this requires us to use an additional relay. We shall call this device a 'RUN relay', and shall use one of the many 'internal relays' (or 'memories' or 'flags' or 'bits') available within the controller for the purpose. Thus our program will feature six sections of ladder diagram, one for each of the controlled items.

Initially we develop our ladder diagram with only plant references (B1, H1, K2, etc.) but in the final solution these references can be substituted by the actual input and output numbers for the chosen controller (as listed in Tables 7.1 and 7.2).

Figure 7.5(*a*) shows the RUN relay control. Internal relay 'RUN' is initiated by a signal from the start switch (S1) but then holds itself on as long as the stop switch (S2) signal is present (S2). When 'RUN' is on it means production is running. The status of 'RUN' will be tested in various parts of the program.

Figure 7.5(*b*) shows the control for valve Y1. With low-level signals for silo 1 (B1) and silo 2 (B3) while production is running ('RUN'), we supply a 1 s pulse to Y1. This prepares silo 1 for filling first.

Figure 7.5(*c*) shows the control for valve Y2. With a high level in silo 1 (B2) while production is running ('RUN'), we supply a 1 s pulse to Y2. This moves the chute toward silo 2.

Figure 7.5(*d*) shows the control for the screw conveyor via K2. The starting impulse is provided by low-level signals from B1 and B3 (production silos empty) and a signal from A1 (product available). Once started, it holds on until both silos are full (i.e. as long as there is no signal from B2 or no signal from B4 and, in addition, production is running ('RUN'), the motor circuit breaker is closed (Q2) and no alarms are present (H1).

Figure 7.5(*e*) shows the rotary discharge valve motor control via K1. It can run if the screw conveyor is already running (K2), has reached full speed (B5), the motor circuit breaker is closed (Q1) and no alarms are present (H1).

Finally, Fig. 7.5(*f*) shows the alarm hooter control (H1). The upper portion shows that if the start switch (S1) is depressed while there is no product available (A1) the hooter is turned on. It then holds itself on until disabled by the stop switch (S2). The lower portion shows that it can also be turned on by either of the motor circuit-breakers tripping (Q1 or Q2) while production is running ('RUN'). The stop switch is still effective for the purpose of turning it off.

Figures 7.6 and 7.7 show how the general solution is translated into the real program for the Mitsubishi and Siemens controllers. In each case the ladder diagram is broken into working sections, the size and arrangement of which is determined by the manufacturer's programming facilities.

In the Mitsubishi ladder diagram, each number on the extreme left refers to the adjacent instruction and indicates its location (step number) in the memory. Thus step 00 holds the

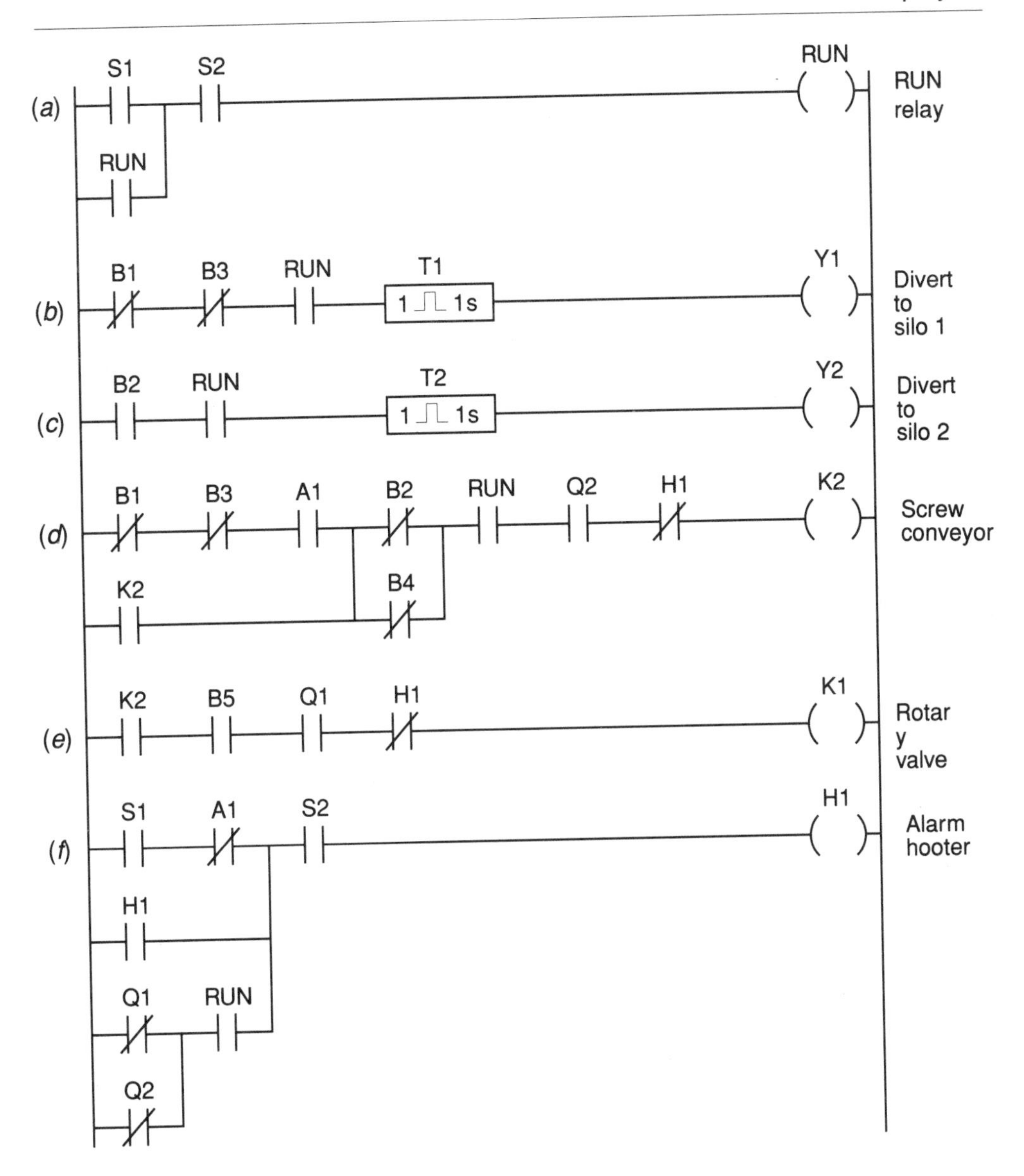

Figure 7.5. Solution in generalized form.

instruction 'LD X410' and step 17 holds the instruction 'LDI X403'. Most instructions require one step but some, such as timers and counters, require two or more. For the F_2 series controller the step numbers are counted in decimal.

The 1 s pulse for driving the solenoid valves is generated by using an ordinary on-delay timer and its inverse state. Thus when signals appear at X404 and M101, T452 starts timing and Y434 is turned on immediately. One second later, T452 turns on and disables Y434. Thus Y434 will have been on for 1 s only.

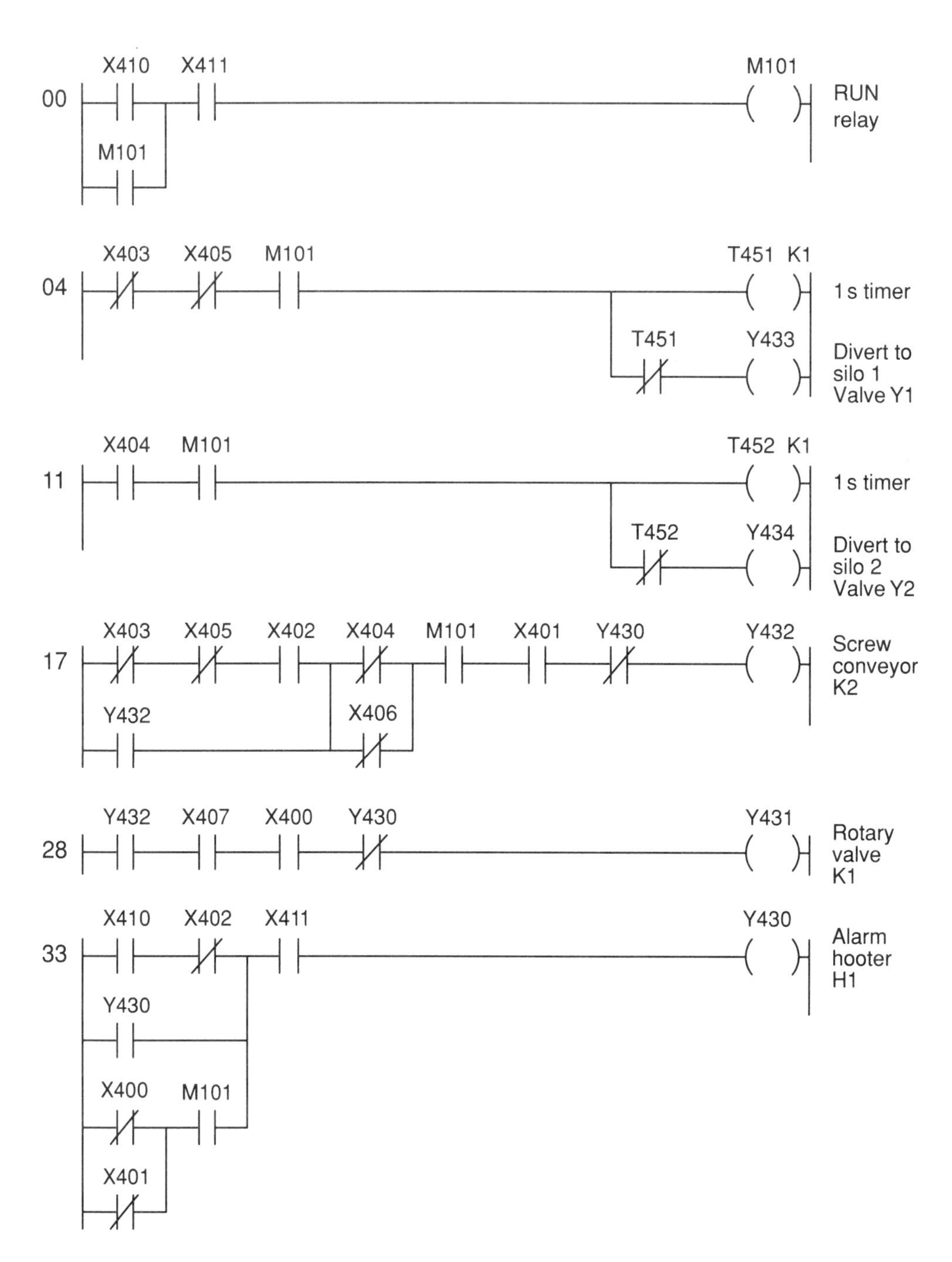

Figure 7.6. Solution for the Mitsubishi F_2-40 controller.

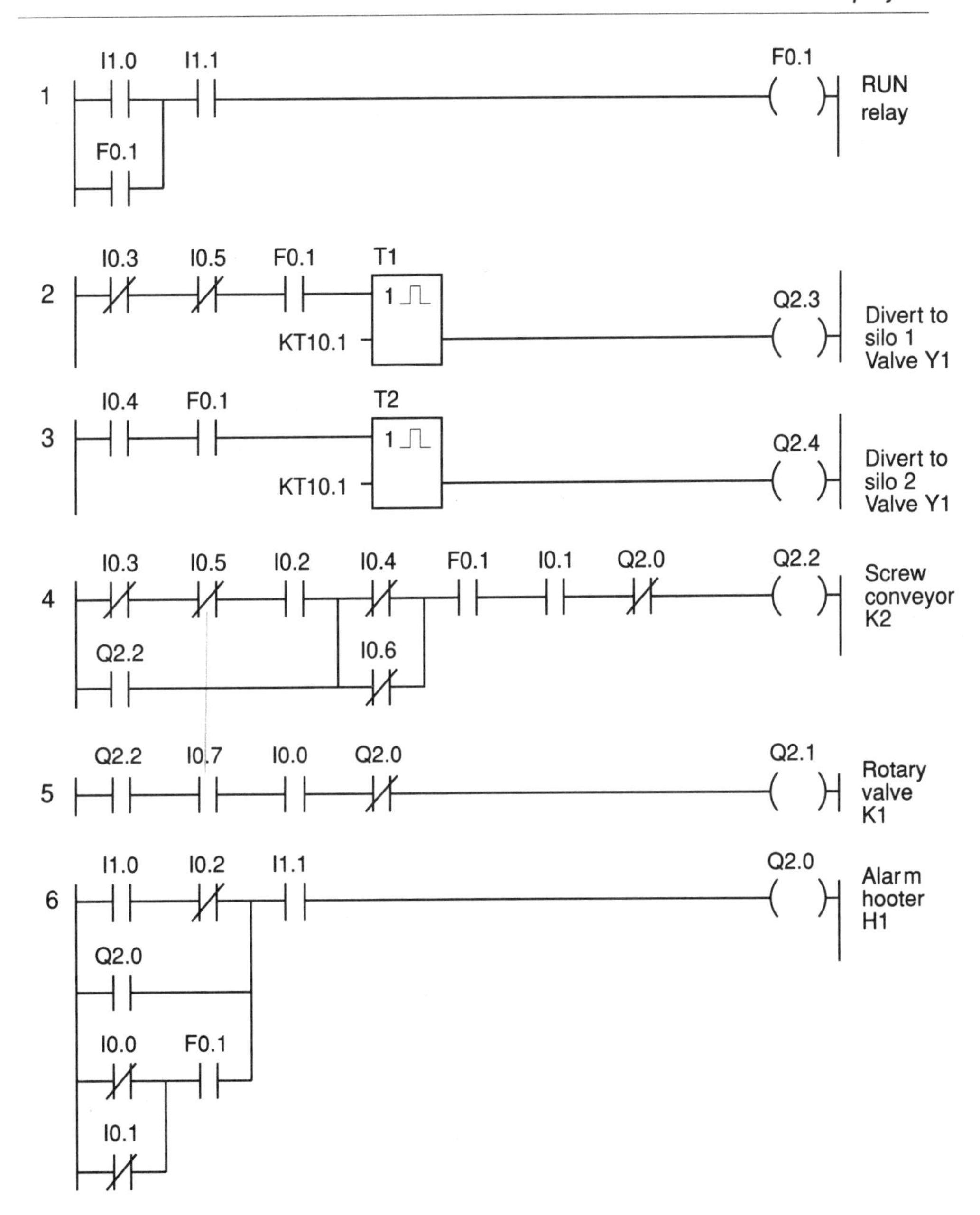

Figure 7.7. Solution for the Siemens SIMATIC® S5-100U controller.

Siemens S5 graphic programming terminals can display the program in three representations—list (STL), ladder (LAD) or logic diagram (CSF) according to the user's preference. In the ladder diagram the numbers on the left are segment numbers. A segment holds all the logic for turning on one output or flag (a flag is an internal relay).

For the Siemens solution we utilized ready-made pulse timers to drive the solenoid valves. Thus at segment 3 we require a signal from I0.4 and F0.1 to generate the 1 s pulse which is then transferred to Q2.4. The notation 'KT10.1' means a fixed time of 10 increments, each of 0.1 s duration, i.e. 10 x 0.1 s = 1 s.

The list programs are given in Table 7.3. They are broadly similar, but the following differences are noteworthy.

In the Mitsubishi case, the list program given is an exact translation of the ladder program shown in Fig. 7.6. Note that the logic beginning at step 17 is complex and requires the use of the ANB instruction to join the two parallel blocks in series. At step 33 we have a different complication: two blocks (each containing AND functions) that have to be joined in parallel. The ORB instruction allows us to do this.

Table 7.3 List programs

Mitsubishi list	Siemens list
LD X410	A I 1.0
OR M101	O. F 0.1
AND X411	A I 1.1
OUT M101	= F 0.1
LDI X403	AN I 0.3
ANI X405	AN I 0.5
AND M101	A F 0.1
OUT T451	LKT 10.1
K 1	SP T 1
ANI T451	A T 1
OUT Y433	= Q 2.3
LD X404	A I 0.4
AND M101	A F 0.1
OUT T452	LKT 10.1
K 1	SP T 2
ANI T452	A T 2
OUT Y434	= Q 2.4
LDI X403	A(
ANI X405	AN I 0.3
AND X402	AN I 0.5
OR Y432	A I 0.2
LDI X404	O. Q 2.2
ORI X406	)
ANB	A(
AND M101	O.N I 0.4
AND X401	O.N I 0.6
ANI Y430	)
OUT Y432	A F 0.1

Table 7.3 List programs (continued)

Mitsubishi list	Siemens list
LD Y432	O.N I 0.0
AND X407	O.N I 0.1
AND X400	)
ANI Y430	A F 0.1
OUT Y431	)
	A I 1.1
LD X410	= Q 2.0
ANI X402	A I 0.1
OR Y430	AN Q 2.0
LDI X400	= Q 2.2
ORI X401	
AND M101	A Q 2.2
ORB	A I 0.7
AND X411	A I 0.0
OUT Y430	AN Q 2.0
	= Q 2.1
	A(
	A I 1.0
	AN I 0.2
	O. Q 2.0
	O
	A(

In the Siemens case, the list (STL) program given is the simplest representation of the ladder diagram shown in Fig. 7.7, but it is not a direct translation. The full translation would require the inclusion of several NOP (no operation) instructions to cater for unused options relating to the timers. The complex logic in segment 4 requires us to use brackets. The instructions situated between 'A(' and ')' are evaluated first, before being joined in series. At segment 6 we use the brackets again to evaluate the logic before the instruction 'AND I1.1'. We also use the O (without the period!) instruction to connect two blocks in parallel.

The step numbers are counted in hexadecimal; each instruction takes one step if displayed on a graphics terminal, two if displayed on a compact list-type terminal.

7.8 Implementation

- The hardware must be purchased, assembled and wired in a suitable panel. This panel must then be located in a safe position in the plant. Drawings must be available.
- The plant wiring must be carried out, the cables being terminated at the input and output terminal racks provided in the panel. Drawings must be available.

- The software must be written to meet the specification and loaded into the programmable controller. Program documentation must be available.
- On completion of the plant wiring, every input and every output device is tested to ensure that it is wired to the correct terminal in the panel and that its signal is generated correctly. A check-list should be used for this purpose and any changes must be recorded.
- Every installation procedure recommended by the programmable controller manufacturer must be followed before powering up.
- The plant must be run on a trial basis. All necessary modifications must be made and recorded.
- A full production run must then be made. All necessary modifications must be made and recorded.
- The final program is stored and a backup copy made (e.g. on EPROM or diskette or cassette tape). Final documentation is made and all drawings are updated.

7.9 Exercises for Chapter 7 (Solutions in Appendix 1)

7.1. Rewrite the ladder and list programs for the Sprecher+Schuh or Telemecanique controllers.

7.2. The project as detailed has a 115 V hooter. Describe all the changes that would be necessary if the hooter was rated for 24 V.

7.3. Using a controller of your choice, devise a suitable layout for the control panel for this project.

7.4. The circuit diagrams show all the level sensing to be done with level switches. What type of sensors might be used instead, and what changes would their use entail?

Chapter 8 Lift project

8.1 Project description

The installation is a simple goods lift operating between two floors. The access door at each floor is opened and closed manually; a mechanical interlock prevents the door from being opened when the lift car is not in place. The car door is also opened and closed manually. The lift is powered by a reversible three-phase motor driving the main pulley through a gearbox. It is fitted with a brake. The installation layout is shown in Fig. 8.1.

The lift can be called to or dispatched from either floor using locally-fitted pushbuttons. Signal lamps on both floors indicate where the lift car is parked and the status of the access doors.

8.2 Equipment

- The lift is to be controlled using a programmable controller.
- The lift motor has a forward (up) contactor K1 and a reverse (down) contactor K2 for switching and a thermal overload relay (F1) for protection.
- The brake Y1 operates in conjunction with the motor. This is arranged electrically and need not involve the programmable controller.
- Signal lamps are assigned as follows:

H1 and H5	'car at ground floor'
H2 and H6	'car at first floor'
H3 and H7	'door open ground floor'
H4 and H8	'door open first floor'

- Alarm bell H9 is fitted to signal some fault conditions.
- Safety limit switches S5 and S6 act as final stopping devices, breaking the control circuit to the motor contactors as well as supplying input signals to the programmable controller. (They are sometimes arranged for more direct control, interrupting the main supply to the motor itself.)
- Calls are initiated by S1 or S3; dispatches are made with S2 or S4.
- Limit switches S9 and S10 detect the lift car parked at ground floor and first floor respectively.
- Limit switches S7 and S8 detect the lift access door closed at ground floor and first floor respectively.
- Limit switch S11 detects the car door closed.

8.3 Specification

1. Under normal circumstances the system will be left in run mode and will operate according to signals from the call and dispatch pushbuttons.
2. The lift is allowed to move under the following conditions:
   - the access doors are closed

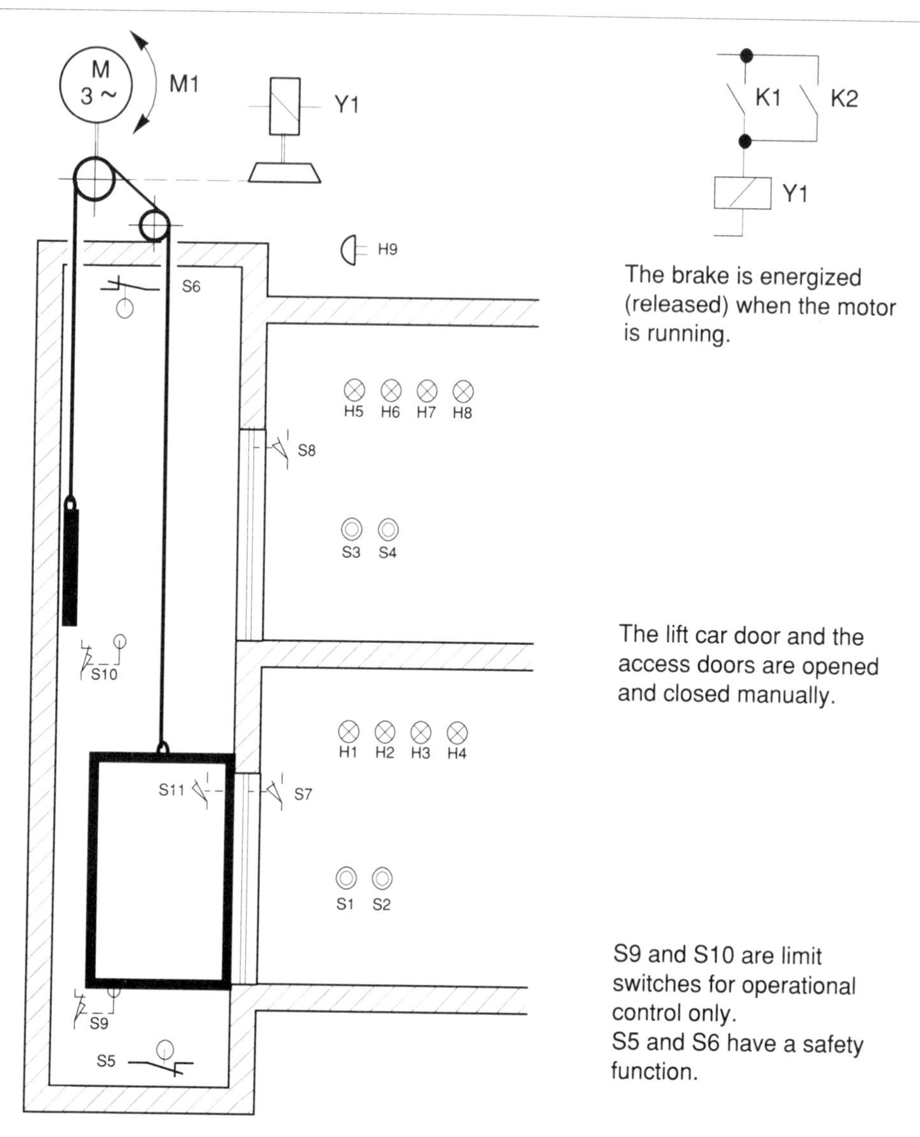

Figure 8.1. Lift installation.

- the car door is closed
- a valid call or dispatch command is given.

3. Once the lift departs from one level it must continue to the other level, regardless of any new commands.
4. The signal lamps illuminate as follows:
   - H1 and H5 when the car is at the ground floor
   - H2 and H6 when the car is at the first floor
   - H3 and H7 when the access door at the ground floor is open
   - H4 and H8 when the access door at the first floor is open.

5. The alarm bell sounds under the following conditions:
   - having departed one level, the lift fails to arrive at the other level within 10 s
   - the lift motor develops an overload.

8.4 Solution

As before, the solution is planned in three steps:

- the main circuit, which is the same whatever controller is used
- the control circuit, including the I/O connections, which is specific to the controller used
- the control program, which is specific to the controller used.

8.5 Main circuit

The main circuit comprises the switchgear and protection for the motor and brake and the necessary power supplies. Figure 8.2 gives details.

8.6 Control circuit

The schedule of inputs we have chosen for this project appears in Table 8.1; the schedule of outputs is given in Table 8.2. The control circuits are developed from these tables. In both cases 24 V d.c. inputs are used.

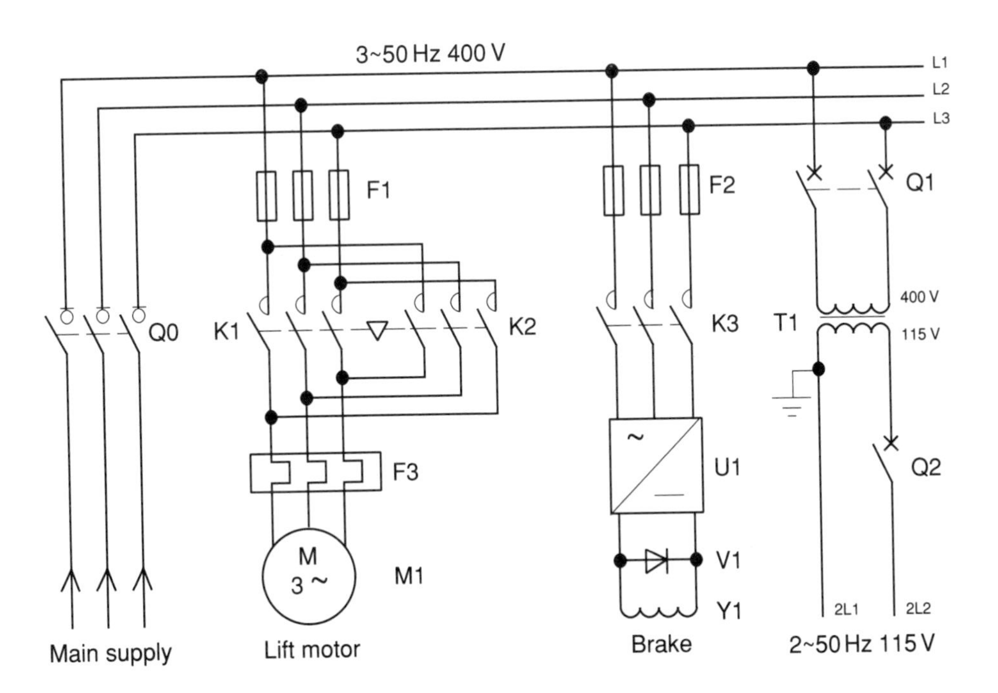

Figure 8.2. Main circuit.

Table 8.1. Schedule of inputs

Device	Item	Input			
		Mitsubishi	Siemens	Sprecher+Schuh	Telemecanique
Call at ground	S1	X401	I0.1	X001	I0,1
Dispatch at ground	S2	X402	I0.2	X002	I0,2
Call at first	S3	X403	I0.3	X003	I0,3
Dispatch at first	S4	X404	I0.4	X004	I0,4
Lower safety limit	S5	X405	I0.5	X005	I0,5
Upper safety limit	S6	X406	I0.6	X006	I0,6
Ground access	S7	X407	I0.7	X007	I0,7
First access	S8	X410	I1.0	X008	I0,8
Lift at ground	S9	X411	I1.1	X009	I0,9
Lift at first	S10	X412	I1.2	X010	I0,10
Lift car door	S11	X413	I1.3	X011	I0,11
Overload relay	F3	X400	I0.0	X012	I0,0

Table 8.2. Schedule of outputs

Device	Item	Output			
		Mitsubishi	Siemens	Sprecher+Schuh	Telemecanique
Raise lift	K1	Y 431	Q2.1	Y001	O0,1
Lower lift	K2	Y 432	Q2.2	Y002	O0,2
'Lift at ground'	H1,5	Y 433	Q2.3	Y003	O0,3
'Lift at first''	H2,6	Y 434	Q2.4	Y004	O0,4
'Ground door'	H3,7	Y 435	Q2.5	Y005	O0,5
'First floor door'	H4,8	Y 436	Q2.6	Y006	O0,6
Alarm bell	H9	Y 437	Q2.7	Y007	O0,7

On the controller, one input is reserved for each switch or contact (a total of 12 inputs). Note that each door switch gives a signal when the door is closed; each level limit switch gives a signal while the lift car is *not* at the corresponding level.

Connection diagrams for the Telemecanique and Sprecher+Schuh controllers are given in Figs 8.3 and 8.4 respectively. Note that signals from safety devices (limit switches and overload relay) may be inputted to the controller; however, the safety *functions* should operate independently of the program (this point is discussed in Chapter 9).

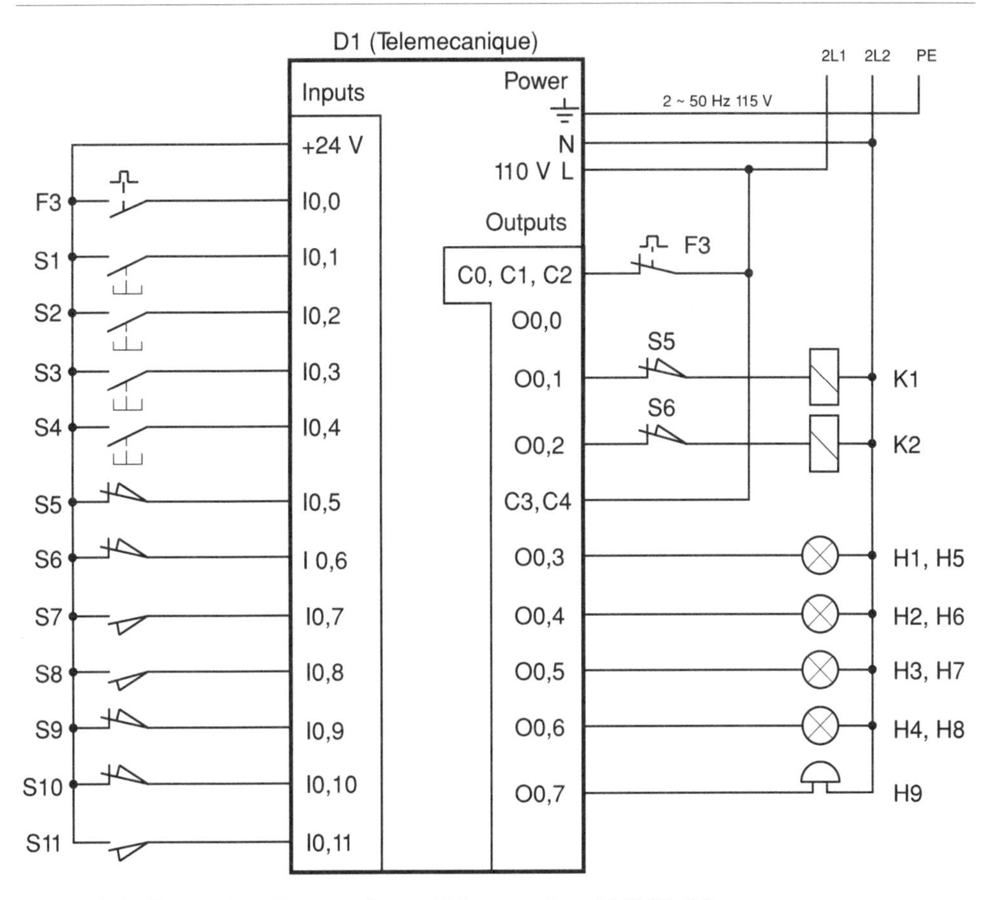

Figure 8.3. Connection diagram for the Telemecanique TSX 17_20.

One output is needed for each motor contactor, for each pair of signal lamps and the bell (a total of seven outputs). Control of the brake contactor K3 is carried out in the wiring as shown in Fig. 8.5.

The outputs serving the motor contactors must incorporate the direct protection of the overload relay. There are two possible approaches to this problem.

One is to route the supply for the 'common' serving the outputs concerned through the overload relay trip contact. This method is illustrated in Fig. 8.3 with the Telemecanique controller.
A second method is to introduce an additional control relay, the coil of which is powered directly through the overload relay trip contact. Make contacts of this new relay are then wired in series with the outputs to be protected. The latter technique is illustrated in Fig. 8.4 with the Sprecher+Schuh controller (the new relay is K4). This approach would normally be adopted if the controller had only one 'common' to serve a large group of outputs.

The bell, signal lamps and contactor coils are all rated for 115 V a.c. for simplicity. It is common practice to supply the signal lamps at 24 V a.c., especially where they are mounted on the door of a control panel (increased risk of direct contact with live parts).

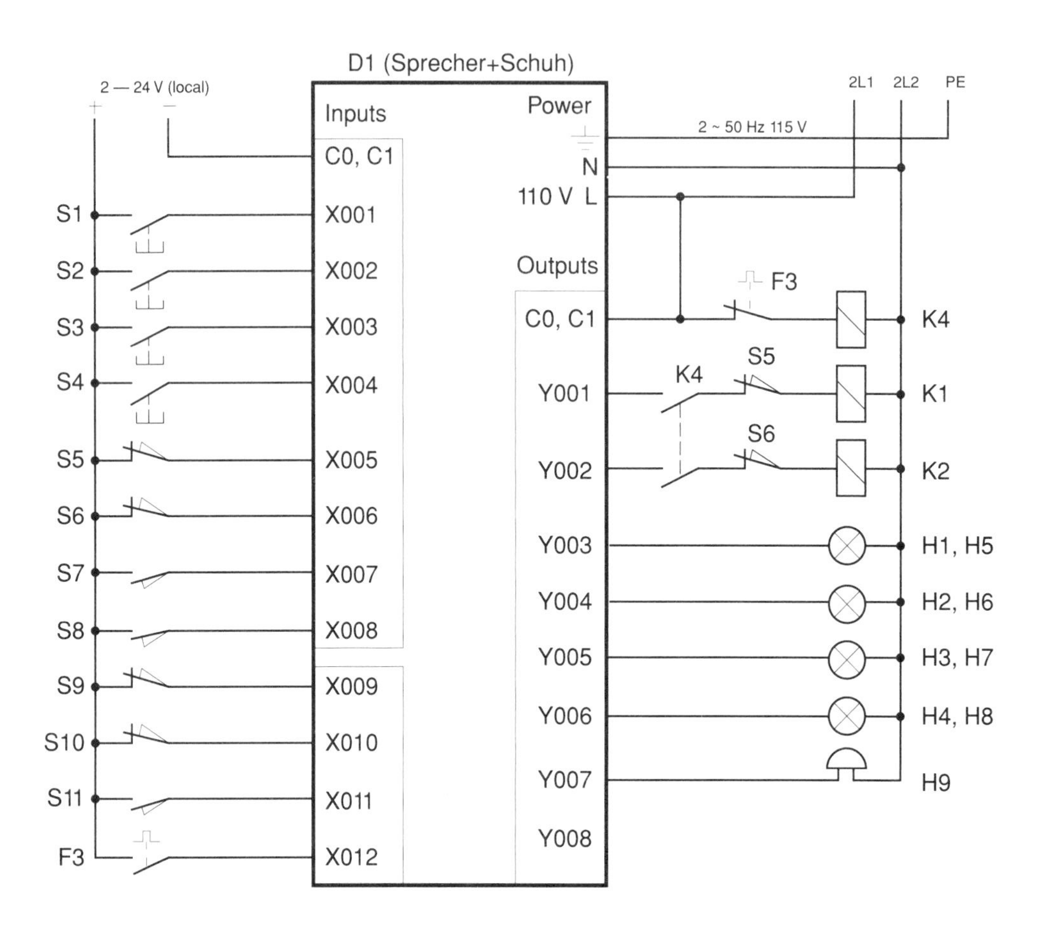

Figure 8.4. Connection diagram for the Sprecher+Schuh SESTEP® 290.

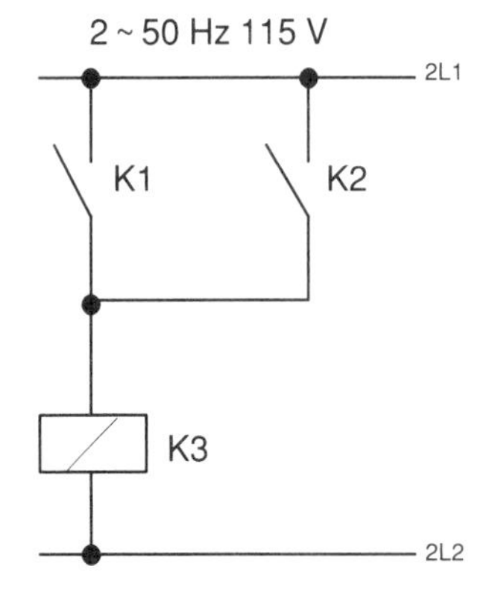

Control of the brake contactor K3

When the lift motor runs, one of the contactors K1 or K2 is energized. An auxiliary contact belonging to one of them switches the supply to the coil of the brake contactor K3 as shown.

This energizes the brake magnet, releasing its hold on the brake shoe.

Figure 8.5. Control of the brake contactor K3.

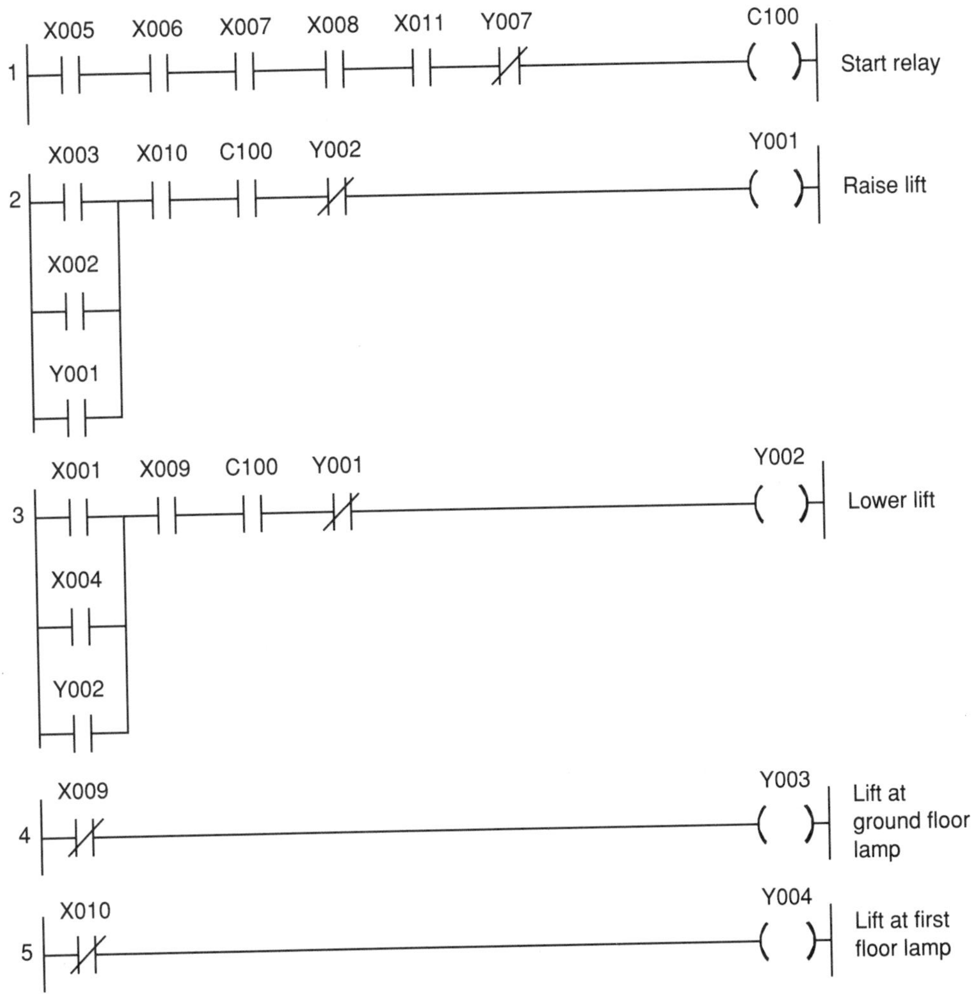

1 X005 X006 X007 X008 X011 Y007 C100 Start relay

2 X003 X010 C100 Y002 Y001 Raise lift
X002
Y001

3 X001 X009 C100 Y001 Y002 Lower lift
X004
Y002

4 X009 Y003 Lift at ground floor lamp

5 X010 Y004 Lift at first floor lamp

Figure 8.6. Sprecher+Schuh ladder solution.

8.7 Program

It is a good policy to develop a general solution first (we did this in the earlier projects, but leave it to the student in this project).

8.7.1 Proposed solution for the Sprecher+Schuh SESTEP® 290 (Figs 8.6 and 8.7)—the corresponding list program is given in Table 8.3

Rung 1

C100 is an internal relay which is turned on when all safety limit switches are closed (signals from X005 and X006) and all doors are closed (signals from X007, X008 and X011) and there is no alarm condition (Y007). C100 is used in rungs 2 and 3 as an enabling signal for the lift motor.

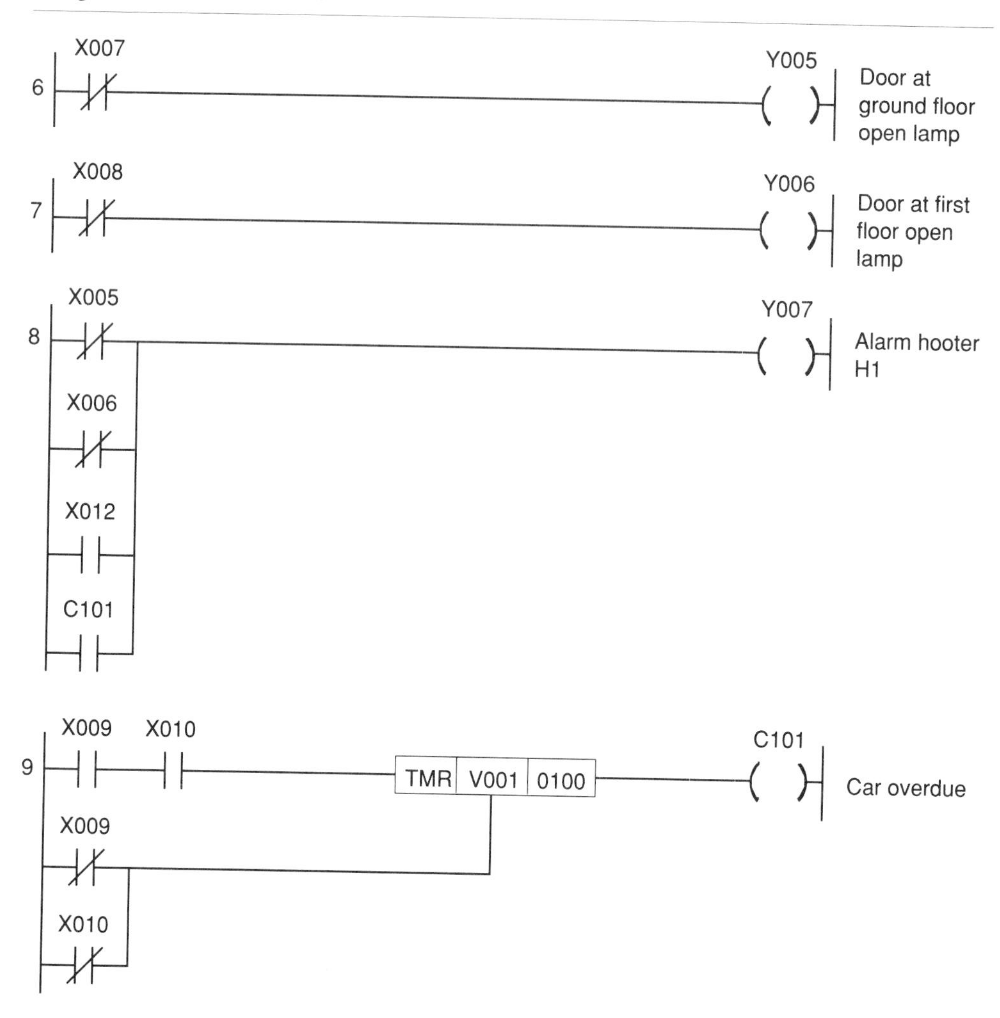

Figure 8.7. Sprecher+Schuh ladder solution (continued).

Rung 2

Y001 is the output for raising the lift. It is started by a signal from X003 (call from first floor) or X002 (dispatch from ground floor). It holds itself on as long as there is a signal from X010 (this will be the case until the lift arrives at the first floor) and from C100 (mentioned in rung 1). Y002 inhibits it if the lift is already descending.

Rung 3

Y002 is the output for lowering the lift. It is started by a signal from X001 (call from ground floor) or X004 (dispatch from first floor). It holds itself on as long as there is a signal from X009 (this will be the case until the lift arrives at the ground floor) and from C100 (mentioned in rung 1). Y001 inhibits it if the lift is already rising.

Rung 4 — Y003 ('lift at ground floor' lamp) is turned on when there is no signal from X009.

Rung 5 — Y004 ('lift at first floor' lamp) is turned on when there is no signal from X010.

Rung 6 — Y005 ('door at ground floor open' lamp) is turned on when there is no signal from X007.

Rung 7 — Y006 ('door at first floor open' lamp) is turned on when there is no signal from X008.

Rung 8 — Y007 is the alarm output. It is used as an inhibiting signal in rung 1 and is turned on by any of the following:

- no signal from X005 (lower safety operated)
- no signal from X006 (upper safety limit)
- a signal from X012 (overload relay operated)
- a signal from C101 (car overdue—see next rung).

Rung 9 — This rung contains the timing for the lift travel. The timer is allowed to run while the lift is between floors (i.e. a signal from X009 and X010). If the preset time (10 s) expires, C101 is turned on to raise the alarm. If the lift arrives at either floor within the time allowed, the absence of a signal from X009 or X010 will reset the timer.

8.7.2 Proposed solution for the Telemecanique TSX 17_20 (Figs 8.8 and 8.9)—the corresponding list program is given in Table 8.3

Label 1 — B100 is an internal relay which is turned on when all safety limit switches are closed (signals from I0,5 and I0,6) and all doors are closed (signals from I0,7, I0,8 and I0,11) and there is no alarm condition (O0,7). B100 is used elsewhere as an enabling signal for the lift motor.

O0,1 is the output for raising the lift. It is started by a signal from I0,3 (call from first floor) or I0,2 (dispatch from ground floor). It holds itself on as long as there is a signal from I0,10 (this will be the case until the lift arrives at the first floor) and from B100 (mentioned above). O0,2 inhibits it if the lift is already descending.

Label 2 — O0,2 is the output for lowering the lift. It is started by a signal from I0,1 (call from ground floor) or I0,4 (despatch from first floor). It holds itself on as long as there is a signal from I0,9 (this will be the case until the lift arrives at the ground floor). O0,1 inhibits if the lift is already ascending.

Label 3 — O0,3 ('lift at ground' lamp) is turned on with no signal from I0,9.

O0,4 ('lift at first' lamp) is turned on with no signal from I0,10.

Table 8.3. List programs

Sprecher+Schuh list		Telemecanique list	
STR	X005	L	I0,5
AND	X006	A	I0,6
AND	X007	A	I0,7
AND	X008	A	I0,8
AND	X011	A	I0,11
AND NOT	Y007	AN	O0,7
OUT	C100	=	B100
STR	X003	L	I0,3
OR	X002	O	I0,2
OR	Y001	O	O0,1
AND	X010	A	I0,10
AND	C100	A	B100
AND NOT	Y002	AN	O0,2
OUT	Y001	=	O0,1
STR	X001	L	I0,1
OR	X004	O	I0,4
OR	Y002	O	O0,2
AND	X009	A	I0,9
AND	C100	A	B100
AND NOT	Y001	AN	O0,1
OUT	Y002	=	O0,2
STR NOT	X009	LN	I0,9
OUT	Y003	=	O0,3
STR NOT	X010	LN	I0,10
OUT	Y004	=	O0,4
STR NOT	X007	LN	I0,7
OUT	Y005	=	O0,5
STR NOT	X008	LN	I0,8
OUT	Y006	=	O0,6
STR NOT	X005	LN	I0,5
OR NOT	X006	ON	I0,6
OR	X012	O	I0,0
OR	Y008	O	B101
OUT	Y007	=	O0,7
STR	X009		
AND	X010	L	I0,9

Table 8.3. List programs (continued)

Sprecher+Schuh list			Telemecanique list	
STR NOT		X009	A	I0,10
OR	NOT	X010	=	T0
TMR		V001	L	T0
		0100	=	B101
OUT		C101		

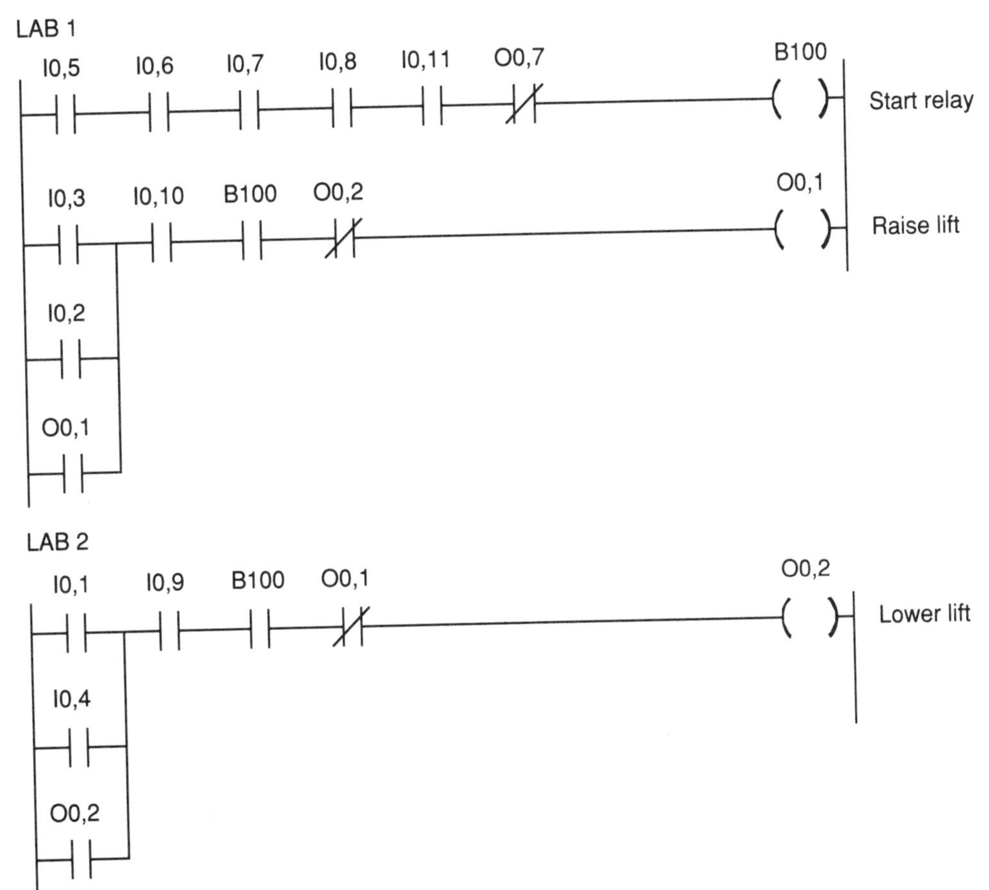

Figure 8.8. Ladder solution on the Telemecanique TSX 17_20.

O0,5 ('ground door open' lamp) is turned on with no signal from I0,7.
O0,6 ('first door open' lamp) is turned on with no signal from I0,8.

Label 4

O0,7 is the alarm output. It is used elsewhere as an enabling signal and is turned on by any one of the following:

- no signal from I0,5 (lower safety operated)

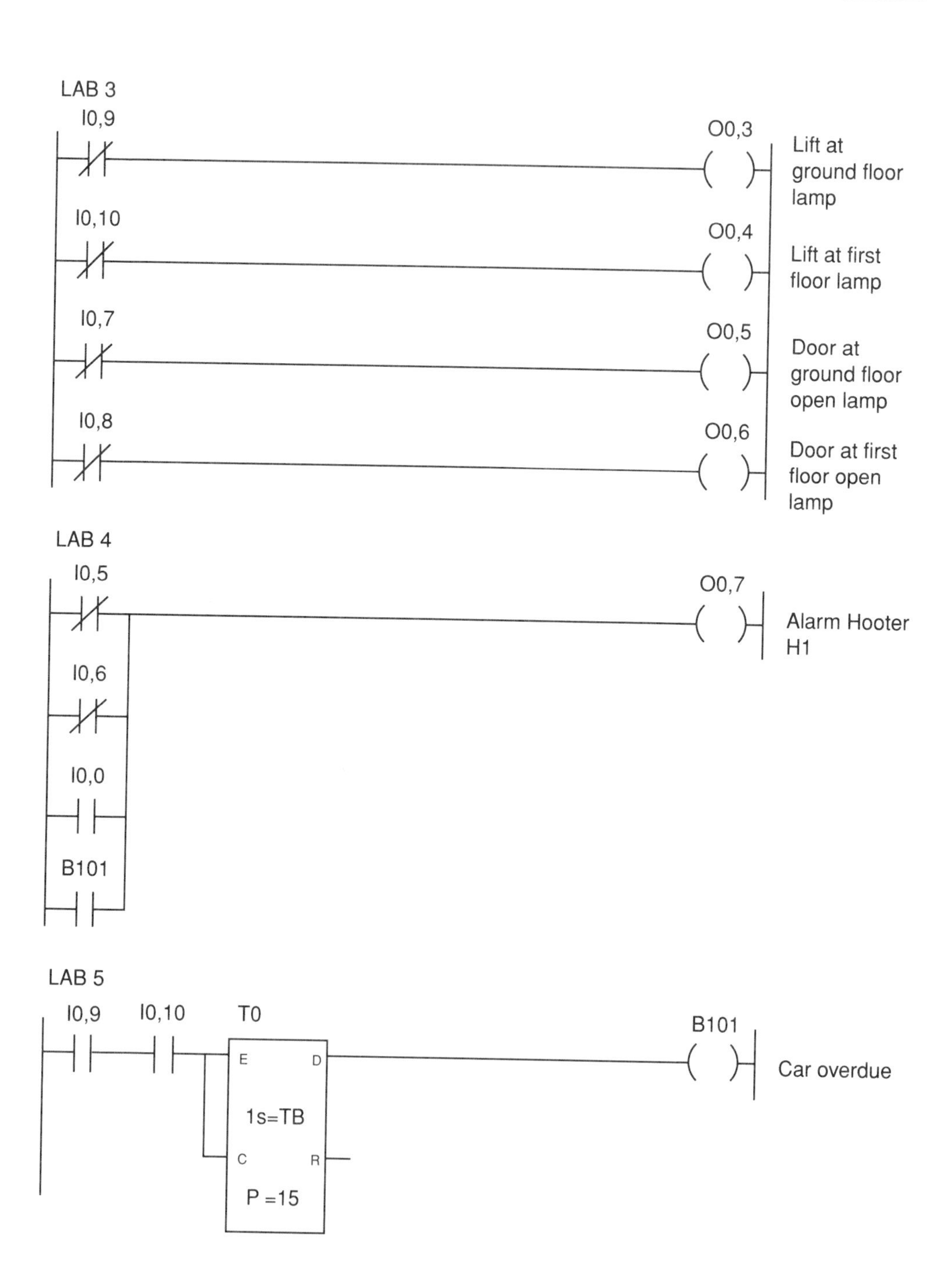

Figure 8.9. Ladder solution on the Telemecanique TSX17_20 (continued).

- no signal from I0,6 (upper safety limit)
- a signal from I0,0 (overload relay operated)
- a signal from B101 (car overdue—see next label).

Label 5

This label contains the timing for the lift travel. Timer T0 is allowed to run while the lift is between floors (i.e. a signal from I0,9 and I0,10). If the preset time (10 s) expires, B101 is turned on to raise the alarm. When the lift arrives at either floor, I0,9 or I0,10 loses its signal and resets T0 once more.

8.8 Implementation

The implementation will be similar to the previous projects'. It is left to the student to write out the main steps as an exercise.

8.9 Exercises for Chapter 8 (Solutions in Appendix 1)

8.1. Modify the control program to provide the following additional facility. When a call or dispatch command is given at one level while the access door at the other level is open, the relevant signal lamps must flash.

8.2. Develop a suitable control program for the Mitsubishi F_2-40 or the Siemens SIMATIC® S5-100U controller.

8.3. Describe all the changes that would be necessary if the signal lamps were rated for 24 V instead of 115 V.

8.4. What happens if the lift is in motion when a power failure occurs?

Note: Some of the exercises require the reader to draw a ladder diagram as part of the solution. This task should be performed with reasonable care, perhaps using a draft ladder programming sheet modelled on Fig. 8.10. This approach will promote clearer draft programs with fewer mistakes and a better record of development work (facilitating easier fault-finding), and is clearly in keeping with the principles of good engineering practice.

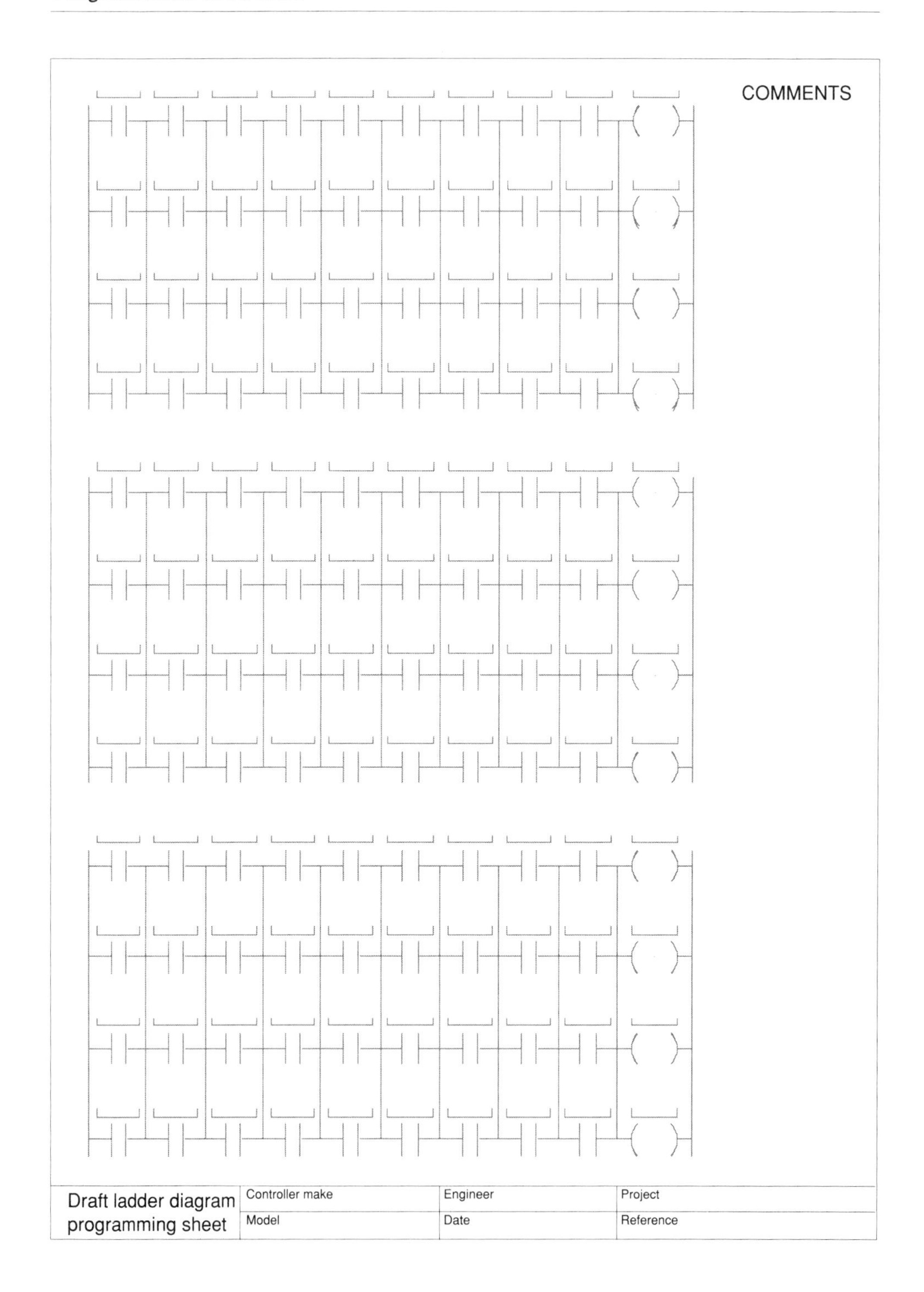

Figure 8.10. Example of a ladder programming sheet suitable for preliminary work.

9.1 Installation

A programmable controller is like any other piece of engineering equipment: installed and used correctly, it will give excellent service. The manufacturer generally gives extensive information on how to install and use the product: it makes sense to follow this advice, if only to ensure that the warranty is preserved. There are a number of points to watch out for.

9.1.1 Physical protection

The programmable controller must be protected against the ingress of water, dust, oils and solvents. These hazards are intolerable to almost any piece of electrical equipment and the normal method of protection is *enclosure*, i.e. fitting the controller in a suitably sealed cabinet. Bear in mind that a higher protective category means less natural ventilation for the controller.

9.1.2 Ambient temperature

For most controllers the specified maximum ambient temperature is 45 to 55°C. This is quite high (ordinary switchgear, for example, must generally be derated for temperatures in excess of 40°C). Temperature problems can arise for any of the following reasons (Fig. 9.1).

(*a*) The controller is located in the same enclosure as other heat sources: the solution is to (a) relocate the controller to a different enclosure or to a low position within the existing one, or (b) improve the ventilation of the existing enclosure, with fan assistance if necessary.

(*b*) There is insufficient space around the controller, resulting in impaired ventilation: the only solution here is to choose an enclosure with more generous dimensions.

(*c*) There is exposure to radiated heat, e.g. direct sunlight: the solution here may be to relocate the whole enclosure or to obscure the heat source.

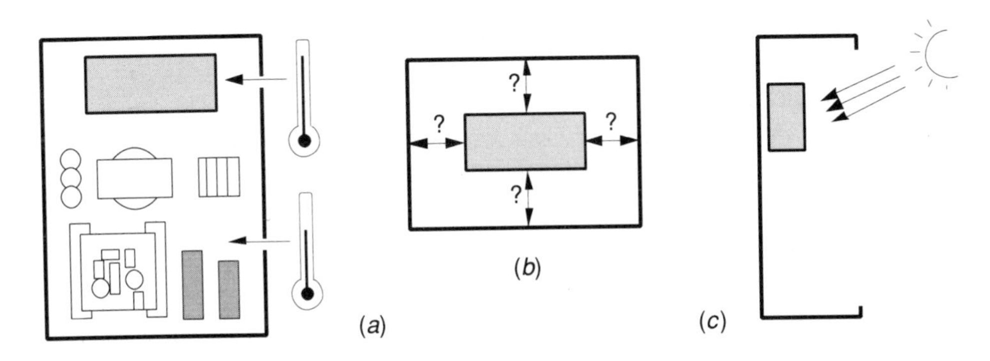

Figure 9.1. Causes of high ambient temperature.

9.1.3 *Electromagnetic interference*

Electromagnetic fields can interfere with the correct operation of the programmable controller. The interference may be from either of the following.

- Electric arcs in heavy switchgear or welding equipment. In this case the interference is radiated through space. The solution is to locate the controller in a steel enclosure, preferably well away from the offending items, and to ensure that its door is kept closed.
- Unsuppressed high-current or high-frequency equipment (e.g. power cables, inverters). In this case the interference can reach the controller through the wiring. The problem usually manifests itself in the form of unwanted input signals, arising for no apparent reason. The solution is to collect the input signals using twisted and screened cables, which are routed well clear of the offending items. Further improvements accrue if the cables are enclosed in steel conduit.

9.1.4 *Wiring*

It is frequently the case that the inputs and outputs operate at different voltages, e.g. 24 V d.c. inputs and 230 V a.c. outputs. There is a natural temptation to choose the I/O cables having regard only for the operating voltage, perhaps forgetting that the input and output cables may share the same duct or trunking inside the enclosure where the controller is mounted. This situation has a recognized potential for danger as well as interference.

The wiring rules in most industrial countries have the same approach to the problem:

1. select *all* the cables for the highest operating voltage present—this gives freedom of choice with cable routes, etc., *or*
2. provide effective separation for different categories of cable—this also helps to reduce interference problems.

9.1.5 *Expansion*

Changes in the size or the character of a plant sometimes means that additional inputs and/or outputs must be introduced. The programmable controller can be extended (up to the limit of the 'max. I/O') by adding extra modules in the case of the rack-type construction, or by adding an expander in the case of the block-type construction. In any event, it is wise to retain the option of expanding by providing the space for it (see Fig. 9.2).

9.1.6 *Connection of field wiring*

The field wiring (the wiring connecting the input and output devices to the control panel) is rarely connected directly to the I/O terminals of the programmable controller because this would entail stripping long lengths of cable sheath and would add to the bulk of cable in the panel trunking. It is more practical to provide terminals at a convenient location (top,

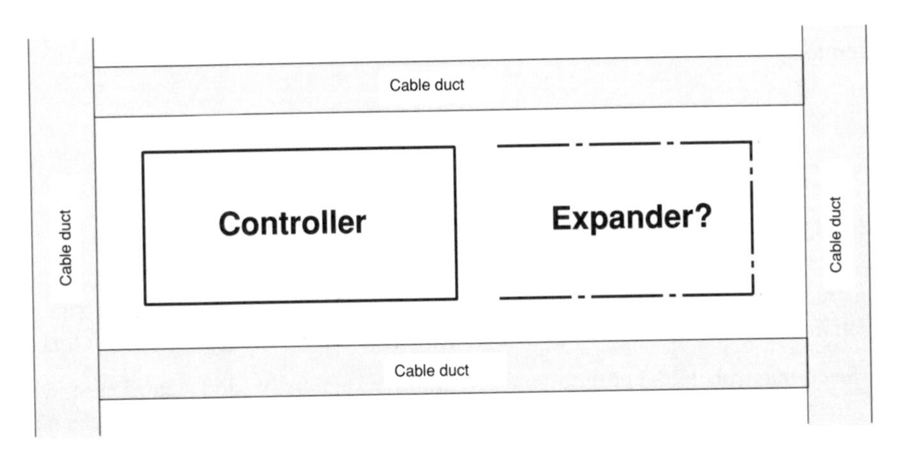

Figure 9.2. Allowing for expansion.

bottom or side of the panel) for the sole purpose of receiving the field wiring. This is clearly convenient from the point of view of connecting the cables, but it is also very useful for fault-finding and testing (Fig. 9.3).

The various terminal racks should be *partitioned*, i.e. the power, input and output terminals should be physically separate and suitably labelled to avoid accidental contact between circuits operating at different voltages. It is also important to allow an adequate number of

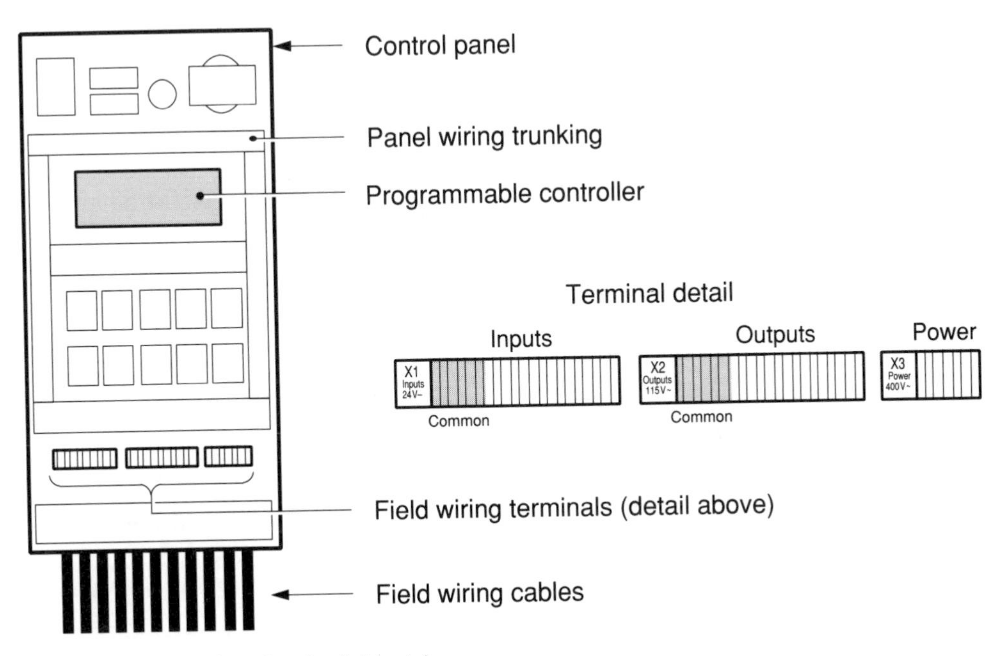

Figure 9.3. Providing for the field wiring.

'common' terminals (L+ and L– for inputs and N for outputs) to avoid multiple connections at one terminal.

9.2 Safety

The construction and operation of the programmable controller is such that safety signals and functions should not be processed through it. There are three main reasons for this.

1. The components in the programmable controller are mostly solid state. These, while extremely reliable, have a different failure pattern to traditional (electromechanical) components. (Of solid state failures, 50 per cent are open-circuit and 50 per cent are short-circuit; of traditional failures, 90 per cent are open-circuit and 10 per cent are short-circuit.) This means that a controller fault, however rare, is as likely to be an incorrect ON signal as it is to be an incorrect OFF signal. In practical terms this means that a non-existent input signal could be seen as being present, or that an output could be wrongly turned ON.
2. At the microcircuit level, the input and output signals are routed along conductors that are separated by tiny fractions of a millimetre. They are thus susceptible to breakdown during transient faults, e.g. due to switching activity or lightning.
3. Within the controller the input and output signals are conveyed to and from the microprocessor along a common set of conductors known as a *bus*. In a typical arrangement 10 conductors manage the affairs of up to 256 input/outputs. This is done by giving each group of inputs or outputs a timed share in the use of the conductors. A failure of the timing has the potential for misdirecting an ON signal, e.g. turning on an output that should be off. There is no comparable failure mechanism in traditional control systems.

The consequences of the above are that we must:

- arrange the emergency and safety functions to be effective even during faults in the programmable controller
- take steps to limit the types of fault that, once imported into the programmable controller, can lead to failure.

9.3 Connection of safety devices

Safety devices are very common in industrial installations, whether or not programmable control is used. Examples are:

- emergency stop switch
- safety limit switch
- protective relay.

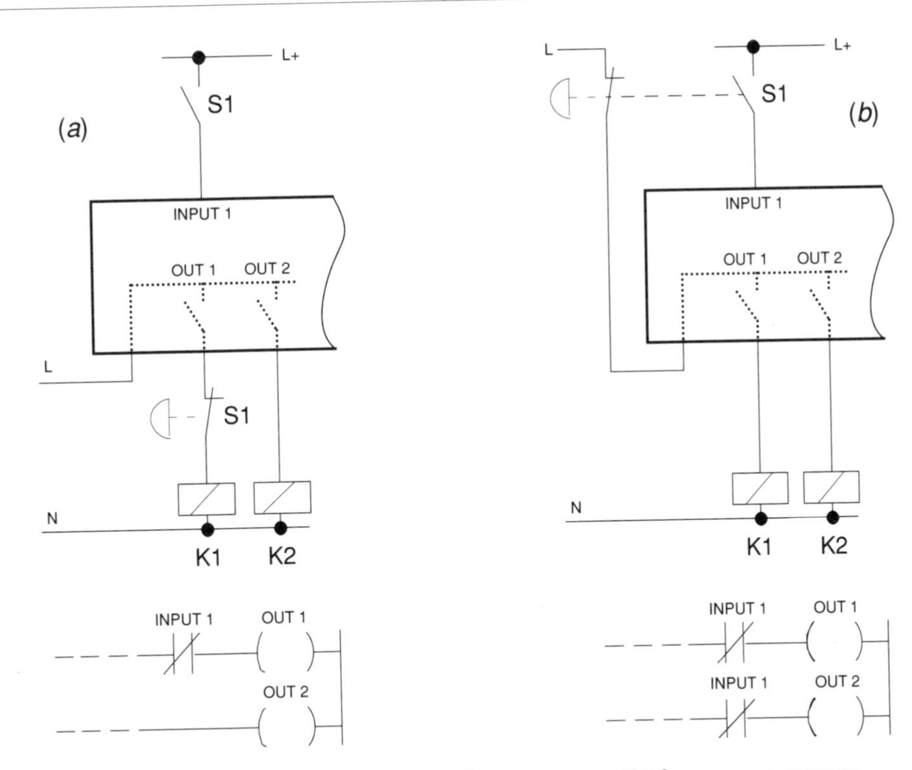

Figure 9.4. Connection of safety devices (a) for one output, (b) for a group outputs.

Figure 9.4 shows how an emergency stop switch is connected in a system that has a programmable controller; this approach is taken with any safety device having a switch-off action. The contact performing the safety function is connected in the *output circuit*. Here, it will remain effective even if an output is incorrectly switched as a result of some controller malfunction. A second contact supplies an input signal which can be used in the program to ensure that the output's 'attempt' to behave correctly. Note that this latter action is straightforward operational control (not safety control).

9.4 Avoiding imported faults

Programmable-controller manufacturers go to a great deal of trouble to ensure that the product is well protected against a variety of electrical hazards. The user can apply a number of measures to ensure that these are supported and not negated. Important points are

- correct earthing procedures
- correct fusing (or circuit breaking)
- overvoltage protection.

Earthing is extremely important. The usual requirement is that a substantial protective conductor be used to connect the earth terminal of the controller to a clean earth. A clean earth is one that is not shared by equipment with very high-rated current or known to cause disturbances. A correctly earthed controller is less susceptible to problems associated with static build-up or to malfunction induced by external earth faults.

Fusing (or circuit-breaking) in the context of overcurrent protection of the output circuits must be done correctly. An overrated fuse or circuit breaker can lead to

- welding of the contacts of an output relay during a short circuit
- destruction of a transistor output due to overcurrent (after which the transistor may remain conductive)
- the destruction of a triac output due to overcurrent (after which the triac may remain conductive).

Overvoltage protection refers to the limitation of damage arising from switching transients or lightning. Protection of individual outputs is dealt with in Section 3.6. However, the potential for trouble can also arise with inputs or even through the power supply. Figure 9.5 shows how voltage-dependent resistors (widely referred to as 'varistors') can be used to dissipate overvoltages harmlessly.

A varistor has a so-called non-linear characteristic. This means it offers virtually infinite resistance as long as the applied voltage is below the critical threshold value. Once the voltage exceeds this value, however, the varistor suddenly loses its resistance and becomes highly conductive. By effectively grounding the circuit, it prevents the transient voltage from penetrating further into the vicinity of the programmable controller.

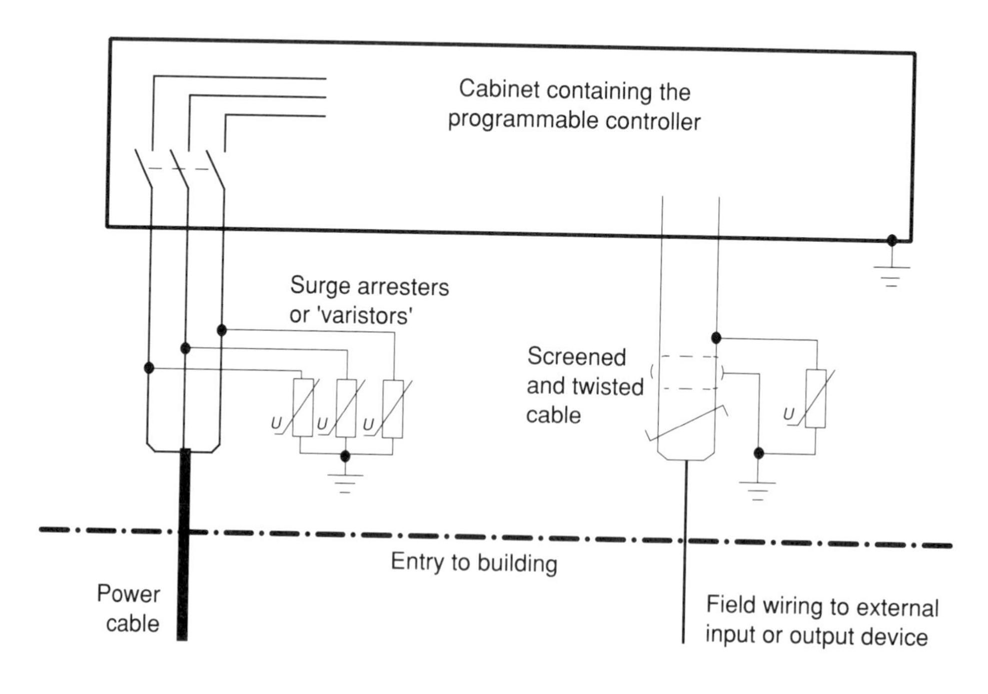

Figure 9.5. Protection against voltage transients and lightning.

9.5 Exercises for Chapter 9 (Solutions in Appendix 1)

9.1. State two physical hazards against which the programmable controller must be protected.

9.2. Input cables operating at 24 V d.c. share the same panel trunking as 230 V a.c. output cables. What voltage rating should the input cables have?

9.3. What is a signal *bus* and why is it important?

9.4. State the general rule to be observed in connecting safety devices to a programmable controller.

Chapter 10 Car-park project

10.1 Project description

A control system is required for a fee-paying car-park with a limited number of places. The entry/exit area of the car-park is shown in Fig. 10.1.

A driver gains access to the car park by depositing a coin in the coin box at the entry. The barrier rises for a preset time to permit entry.

The exit barrier rises automatically for a preset time when a departing car approaches it.

10.2 Equipment

- Entry and exit barriers are pneumatically operated.
- Illuminated signs indicate whether the car-park has places or is full.
- A coin-operated switch is placed 'in-route' to the entry barrier.
- An inductive vehicle sensor is fitted 'in-route' to the exit barrier.
- Each barrier is fitted with an 'up' and a 'down' proximity sensor.
- A text display situated in the attendant's office supplies status information.

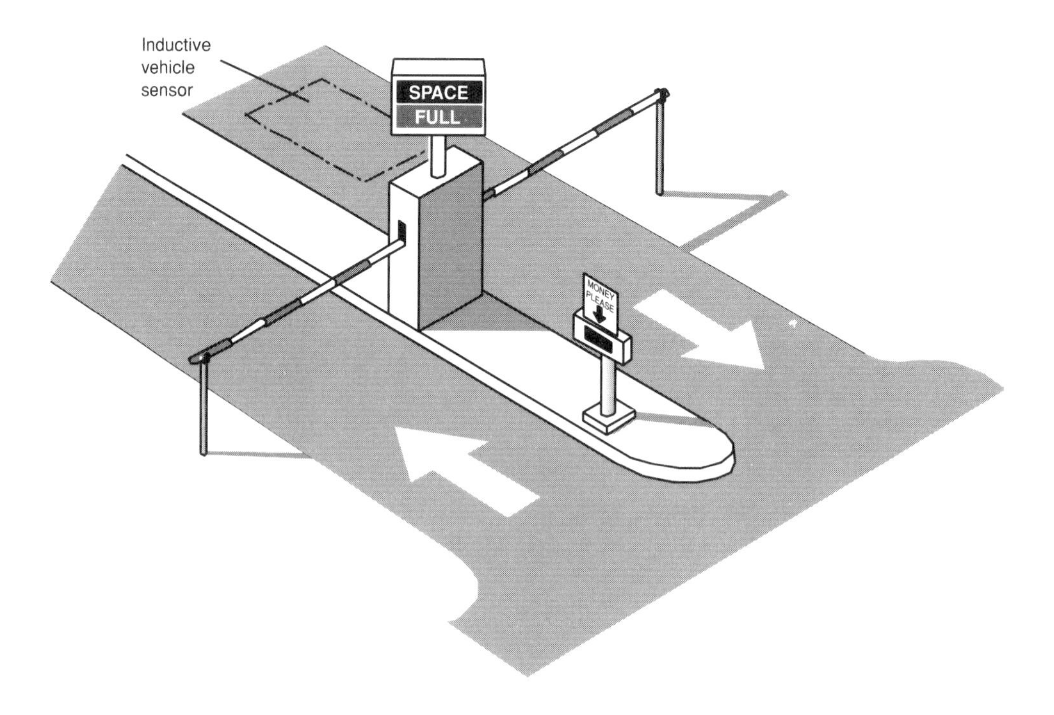

Figure 10.1. Car-park with automatic entry/exit.

10.3 Specification

1. Control is to be entirely automatic; the number of cars in the car-park is to be limited to 50. A fixed fee applies to all entrants.
2. On payment of the correct fee, the entry barrier must rise and remain open for 6 s. It must then drop again until the next fee is paid.
3. When a departing car approaches the exit barrier, it passes over the inductive vehicle sensor. The exit barrier must rise and remain open for 6 s. It must then drop automatically until the next departing car approaches.
4. When space is available, the 'SPACE' indicator is to operate.
5. When the car-park is full, the 'FULL' indicator is to operate and no more cars are to be allowed to enter.
6. The text display is to provide for up to 32 messages. Message 1 indicates the number of cars remaining in the car-park. Other messages are to be devised later.

10.4 Solution

The solution is presented in four steps:

- an explanation of the pneumatic circuit, which is the same regardless of the controller used
- the control circuit including input and output connections, which is specific to the controller used
- the control program for the entry/exit, which is specific to the controller used
- the functioning and control of the text display.

10.5 Pneumatic circuit

For those unfamiliar with pneumatics, the following description may be helpful. The movement of each barrier depends on a cylinder driven by compressed air (see Fig. 10.2). In a typical arrangement the cylinder is fitted with a three-port solenoid valve at each end and can be made to extend or retract by energizing one of these valves. Details of the action of the cylinder driving the entry barrier are shown in Fig. 10.3.

Unoperated	The two sides of the cylinder are vented and neither solenoid valve is energized.
Extend	Valve Y1 is energized; air enters the left side of the cylinder and drives the piston forward (the right side is still vented because Y2 is de-energized).
Retract	Valve Y1 is de-energized to vent the left side of the cylinder. Valve Y2 is energized; air enters the right side of the cylinder and drives the piston back.

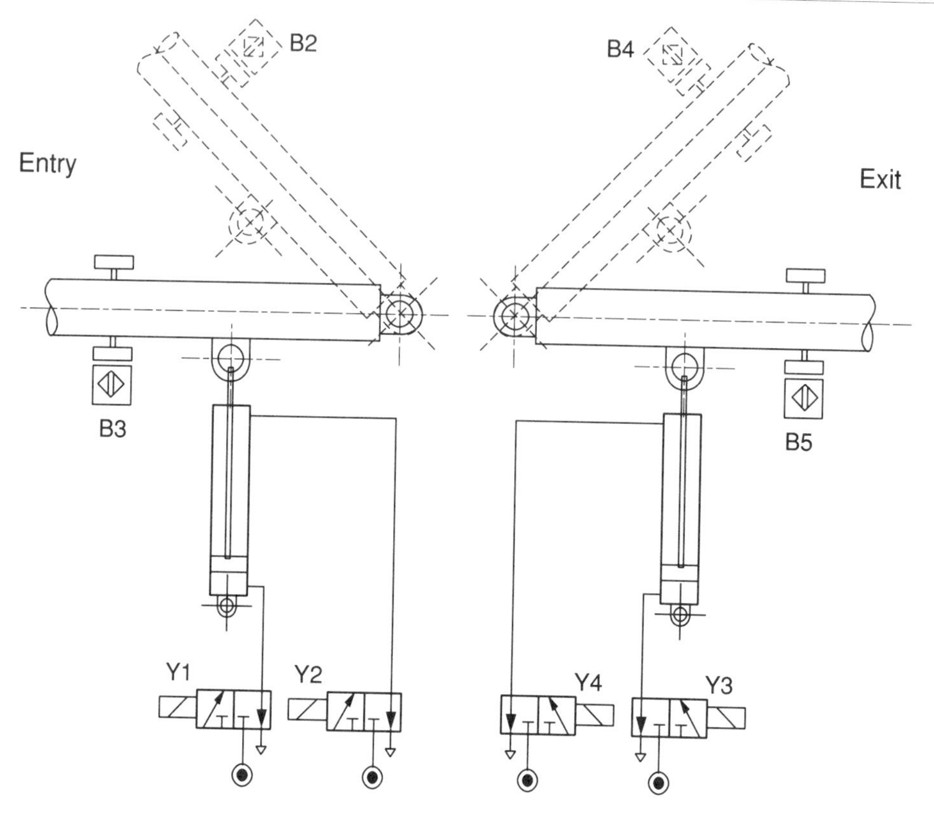

Figure 10.2. Pneumatic circuit for the barriers.

The symbols in Fig. 10.3 comply with ISO 1219. The valves are shown in the unoperated position. The operated position is visualized by mentally shifting the whole 'block' so that its other half is aligned with the ports.

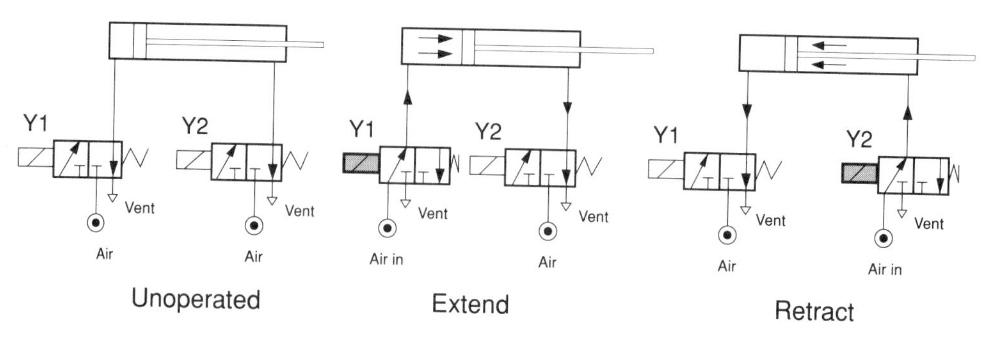

Figure 10.3. Valve and cylinder operation.

Table 10.1. Schedule of inputs

Device	Item	Input			
		Mitsubishi	Siemens	Sprecher+Schuh	Telemecanique
Coin switch	S1	X400	I0.0	X001	I0,0
Vehicle sensor	B1	X401	I0.1	X002	I0,1
Entry barrier up	B2	X402	I0.2	X003	I0,2
Entry barrier down	B3	X403	I0.3	X004	I0,3
Exit barrier up	B4	X404	I0.4	X005	I0,4
Exit barrier down	B5	X405	I0.5	X006	I0,5

Table 10.2. Schedule of outputs

Device	Item	Output			
		Mitsubishi	Siemens	Sprecher+Schuh	Telemecanique
Raise entry barrier	Y1	Y430	Q2.0	Y001	O0,0
Lower entry barrier	Y2	Y431	Q2.1	Y002	O0,1
Raise exit barrier	Y3	Y432	Q2.2	Y003	O0,2
Lower exit barrier	Y4	Y433	Q2.3	Y004	O0,3
'Space' lamp	H1	Y434	Q2.4	Y005	O0,4
'Full' lamp	H2	Y435	Q2.5	Y006	O0,5
Text display	H3	Y530..535	Q3.0..3.5	Y009..14	O1,0..1,5

10.6 Control circuit

The I/O allocation is detailed in Tables 10.1 and 10.2. The control circuits shown in Figs 10.4 (for Siemens) and 10.5 (for Telemecanique) are based on these tables.

Each proximity sensor is a three-wire, PNP type, (a so-called 'source input'), and adds approximately 10 mA to the load on the d.c. power supply. For this project we assume it gives an input signal to the programmable controller when the target is nearby—note, however, that a device with the opposite switching pattern is also available.

The coin switch generates a brief but adequate signal when a valid fee is paid.

The inductive vehicle sensor (a sub-surface wire loop) is used in conjunction with a suitable amplifier which generates an input signal when the vehicle is detected.

The solenoid valves and sign lamps are all rated for 115 V a.c. Note, however, that 24 V and 48 V supplies are also quite common for valves; if either of these voltages were to be chosen for this project we would have to arrange the output wiring with some caution.

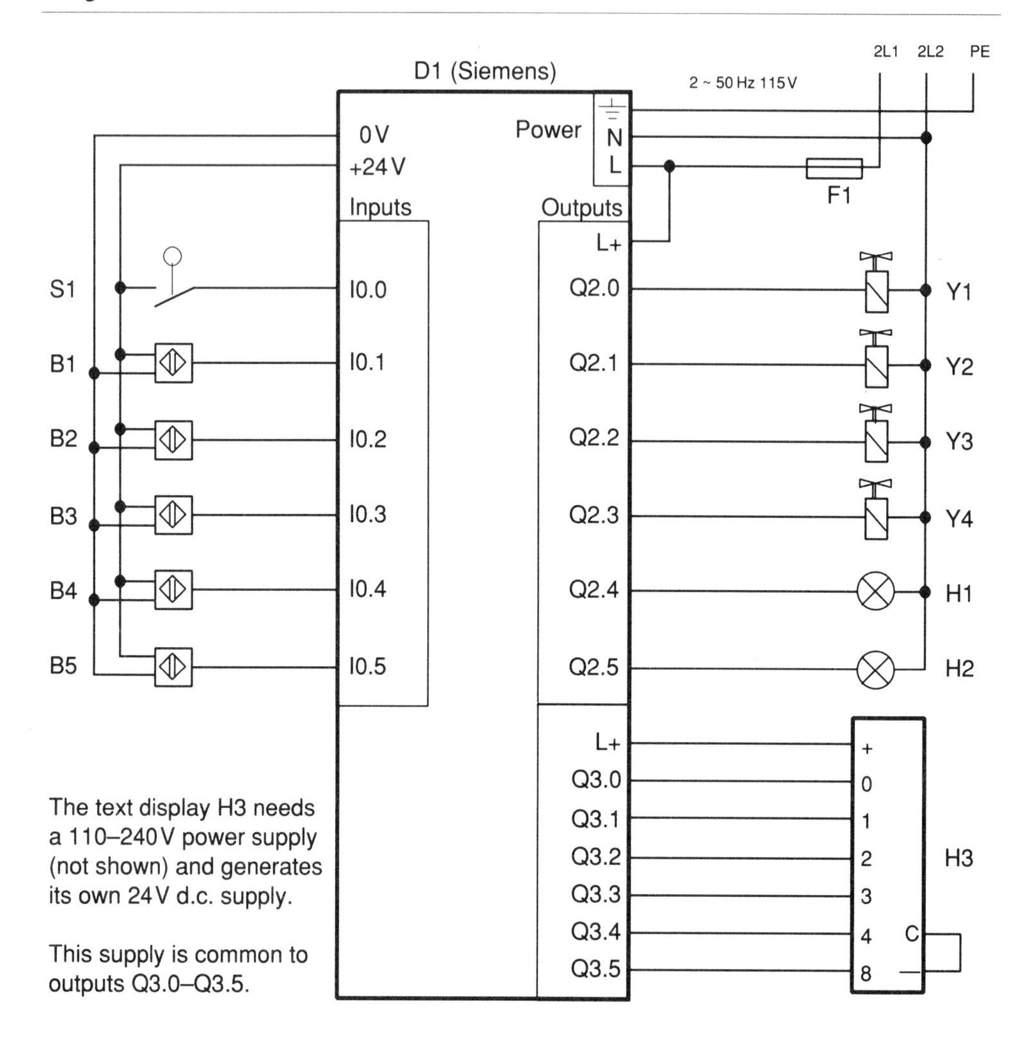

Figure 10.4. Connection diagram for the Siemens SIMATIC® S5-100U.

Specifically, we would have to ensure the separation of the 115 V supply and the valve supply. Also, transient protection (an *RC* network or a varistor fitted to the coil) is appropriate for each valve but has been omitted from our circuit diagram for clarity. Each valve is non-latching and must be energized for the full period of its action.

To drive the text display, we need six transistor outputs (explained in Section 10.8) and a 24 V d.c. supply. In the Siemens case, module 3 on the controller has transistor outputs so we can connect our display to these outputs. In the Telemecanique case the existing outputs are of the relay type and it is necessary to fit an extension block (type TSX DSF 612) having transistor outputs. The 24 V supply can be taken from the display or from the programmable controller—we have chosen the display's supply.

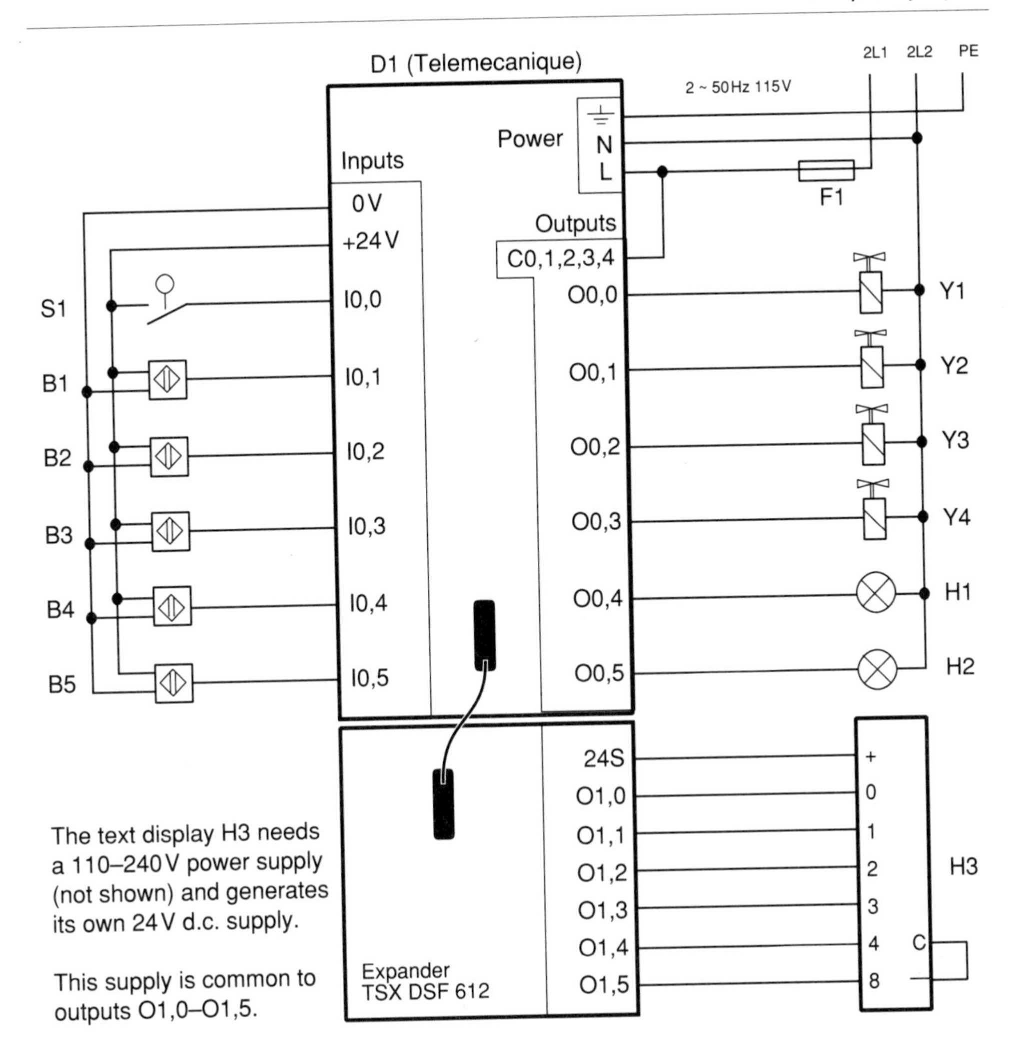

Figure 10.5. Connection diagram for the Telemecanique TSX 17_20.

10.7 Program for entry/exit control

Solutions are offered for the Siemens and Telemecanique controllers. The two programs are very similar except for the counter implementation. We have used a count value of 5 instead of 50 purely for convenience of testing. The list programs are left for completion by the student.

10.7.1 Proposed solution for Siemens SIMATIC® S5-100U (Figs 10.6 and 10.7)

Segment 1

Q2.0 is the output for raising the entry barrier. A signal from the coin box I0.0 turns it on; it holds itself on until internal relay F0.1 gives a signal (this happens when the barrier has been raised for the pre-set time). Two inhibiting signals are included, namely Q2.5 (car-park full) and Q2.1 (lower entry barrier).

Segment 2

Q2.1 is the output for lowering the entry barrier. Internal relay F0.1 turns it on (up-time expired); it holds itself on until I0.3 gives a signal (this happens when the barrier is down). Q2.0 is an inhibiting signal (raise entry barrier).

Segment 3

I0.2 gives a signal when the entry barrier is fully up. This starts on-delay timer T1. After 6 s internal relay F0.1 is turned on.

Segment 4

Q2.2 is the output for raising the exit barrier. A signal from the vehicle sensor I0.1 turns it on; it holds itself on until internal relay F0.2 gives a signal (this happens when the barrier has been raised for the pre-set time). Q2.3 is an inhibiting signal (lower exit barrier).

Segment 5

Q2.3 is the output for lowering the exit barrier. Internal relay F0.2 turns it on (up-time expired); it holds itself on until I0.5 gives a signal (this happens when the barrier is down). Q2.2 is an inhibiting signal (raise exit barrier).

Segment 6

I0.4 gives a signal when the exit barrier is fully up. This starts on-delay timer T2. After 6 s internal relay F0.2 is turned on.

Segment 7

C1 is the vehicle counter. Note that the Siemens counter gives an output signal *for any number except zero*. Our strategy has been to load the count value with an unused internal relay F127.7. This sets the count at 5. The output Q2.4 (SPACE lamp) is turned on immediately. Each new signal from I0.2 (entry barrier up) causes the counter to count down. When five cars have passed, the counter signal to Q2.4 is turned off. Each new signal from I0.4 (exit barrier up) counts it up again and restores the signal to Q2.4.

Segment 8

Q2.5 (FULL lamp) is turned on whenever Q2.4 (SPACE lamp) is not on. Note that Q2.5 is used in segment 1 to inhibit the raising of the entry barrier when the car-park is full.

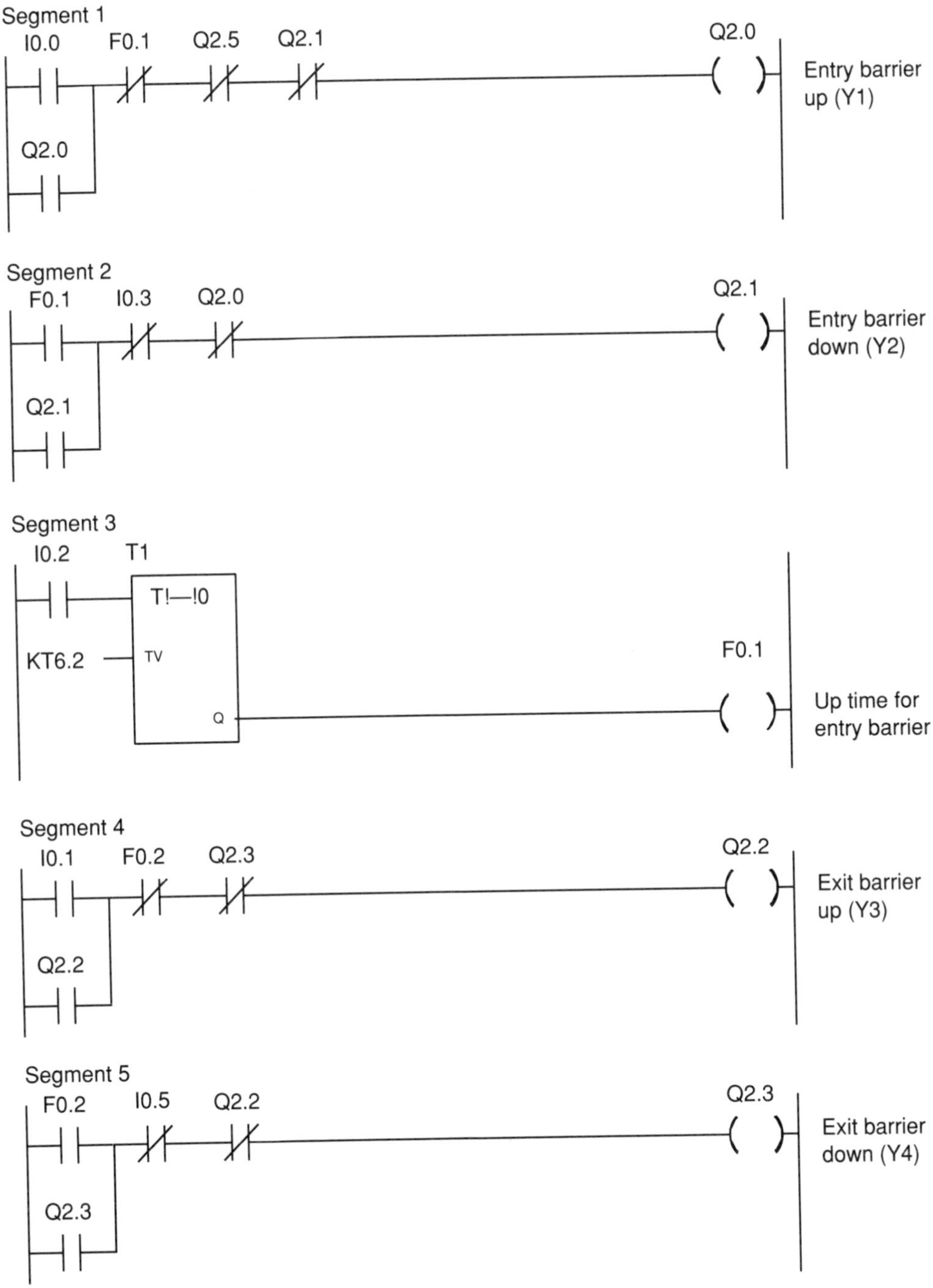

Segment 1
I0.0 F0.1 Q2.5 Q2.1 Q2.0
Entry barrier up (Y1)
Q2.0

Segment 2
F0.1 I0.3 Q2.0 Q2.1
Entry barrier down (Y2)
Q2.1

Segment 3
I0.2 T1
T!—!0
KT6.2 TV
Q
F0.1
Up time for entry barrier

Segment 4
I0.1 F0.2 Q2.3 Q2.2
Exit barrier up (Y3)
Q2.2

Segment 5
F0.2 I0.5 Q2.2 Q2.3
Exit barrier down (Y4)
Q2.3

Figure 10.6. Ladder solution for Siemens SIMATIC® S5-100U.

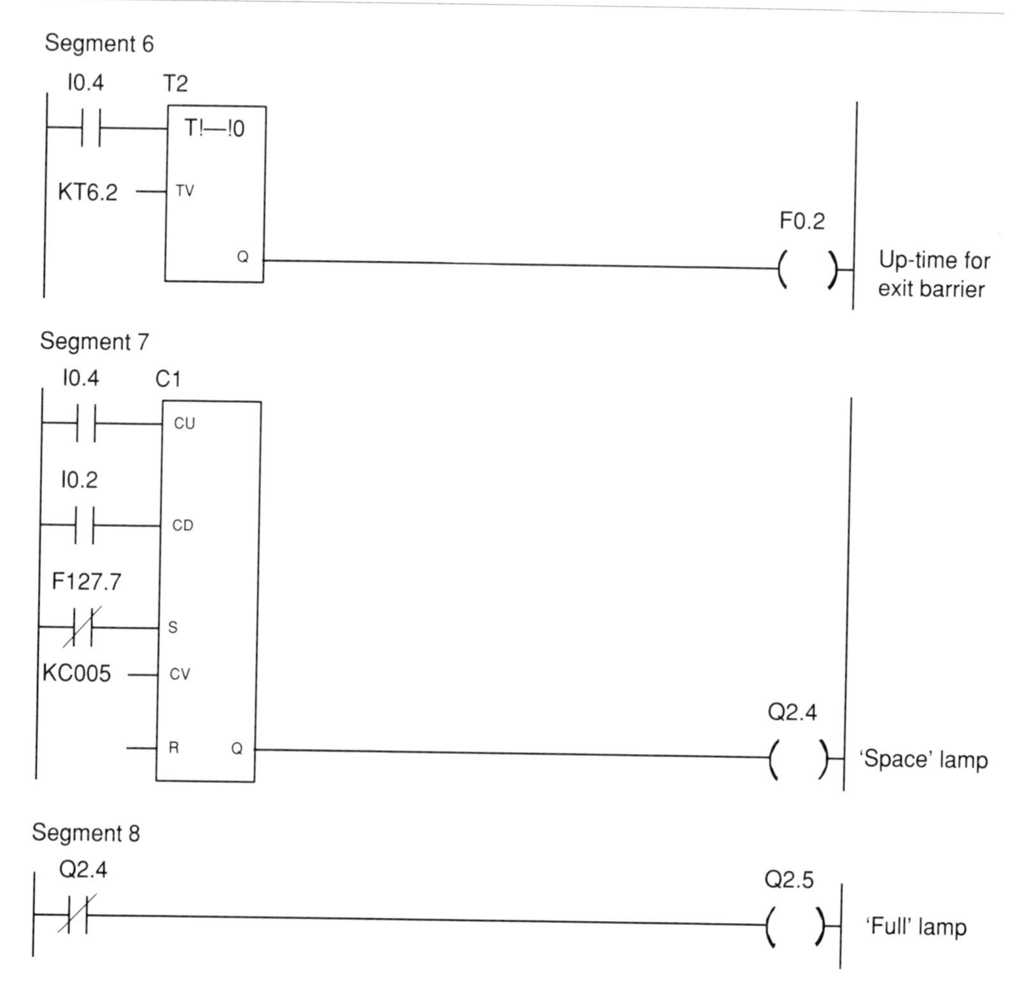

Figure 10.7. Ladder solution for Siemens SIMATIC® S5-100U (continued).

10.7.2 Proposed solution for Telemecanique TSX 17_20 (Figs 10.8 and 10.9)

Label 1

O0,0 is the output for raising the entry barrier. A signal from the coin box I0,0 turns it on; it holds itself on until internal relay B1 gives a signal (this happens when the barrier has been raised for the pre-set time). Two inhibiting signals are included, namely O0,5 (car-park full) and O0,1 (lower entry barrier).
O0,1 is the output for lowering the entry barrier. Internal relay B1 turns it on (up-time expired); it holds itself on until I0,3 gives a signal (this happens when the barrier is down). O0,0 is an inhibiting signal (raise entry barrier).

Label 2

I0,2 gives a signal when the entry barrier is fully up. This starts on-delay timer T1. After 6 s internal relay B1 is turned on.

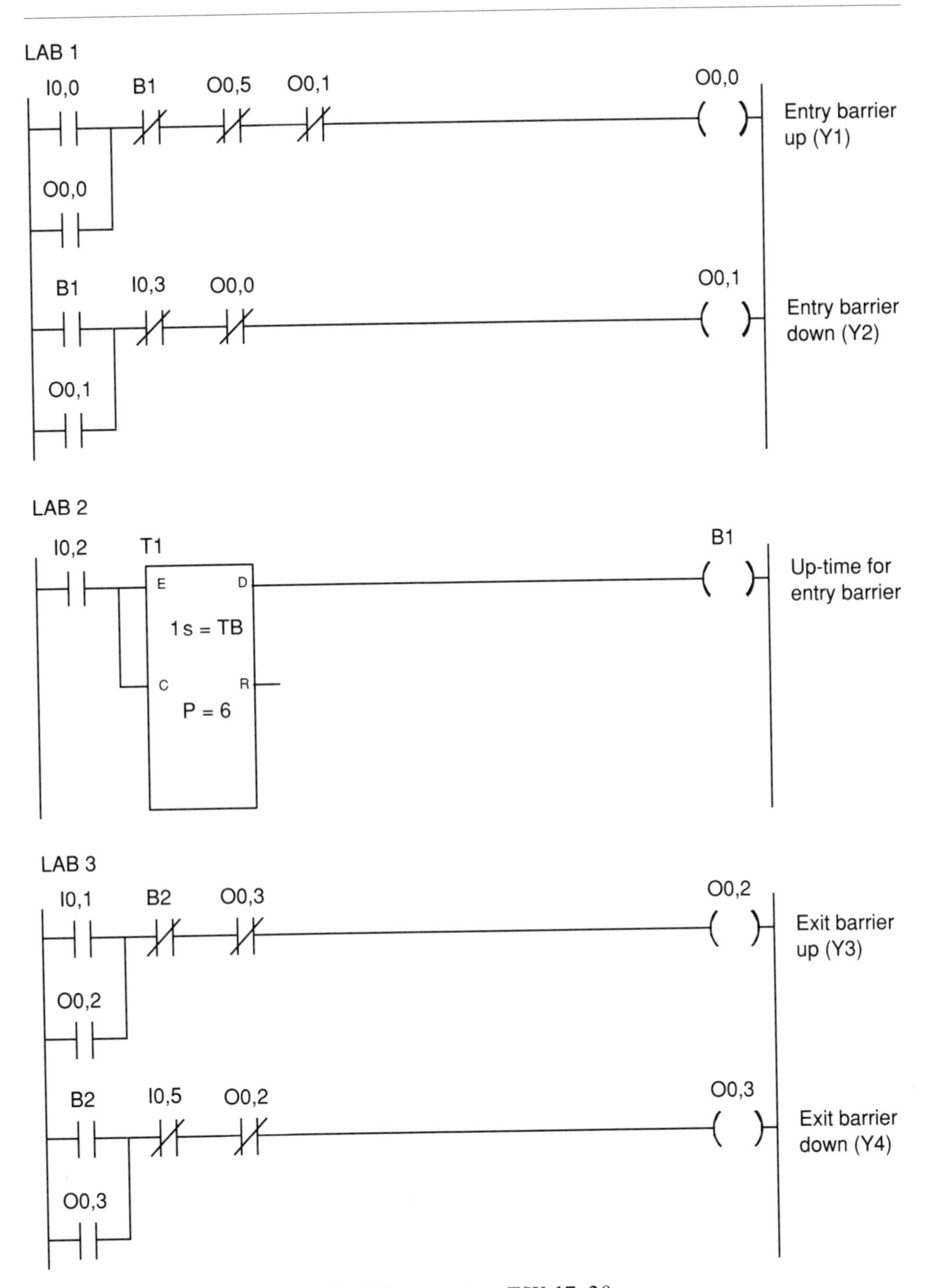

Figure 10.8. Ladder solution for Telemecanique TSX 17_20.

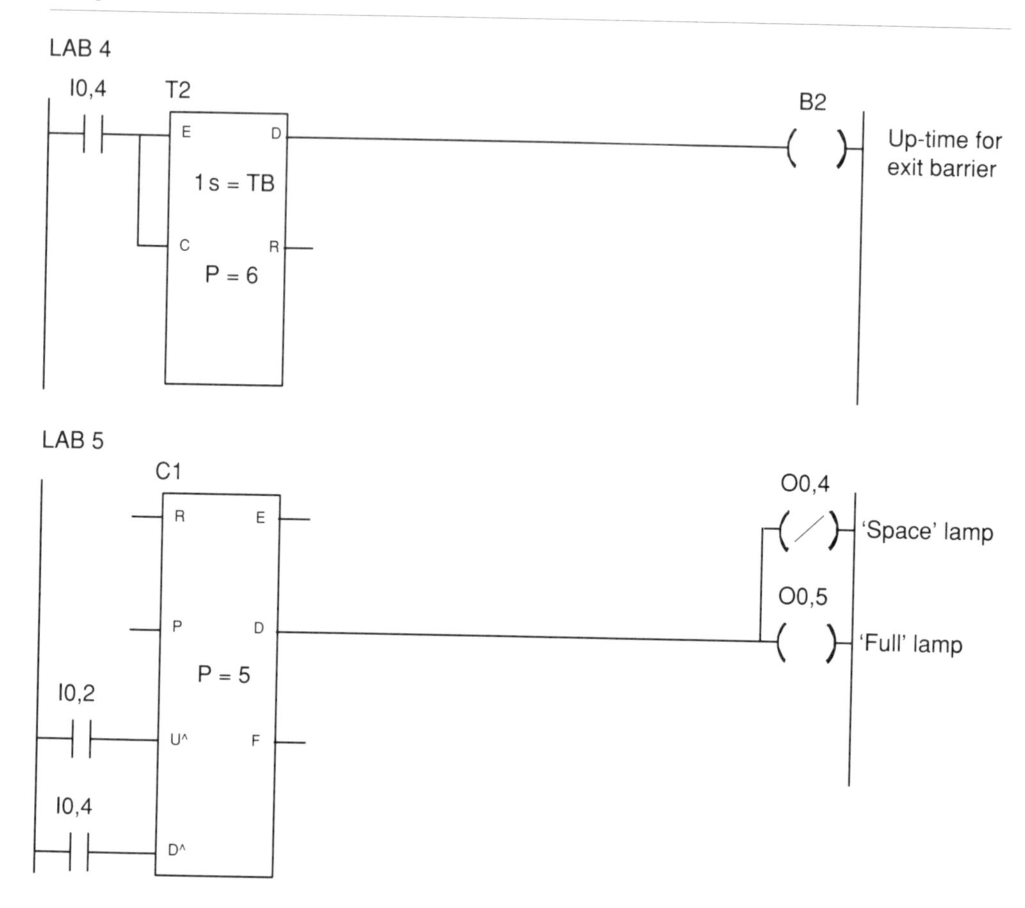

Figure 10.9. Ladder solution for Telemecanique TSX 17_20 (continued).

Label 3 — O0,2 is the output for raising the exit barrier. A signal from the vehicle sensor I0,1 turns it on; it holds itself on until internal relay B2 gives a signal (this happens when the barrier has been raised for the pre-set time). O0,3 is an inhibiting signal (lower exit barrier).
O0,3 is the output for lowering the exit barrier. Internal relay B2 turns it on (up-time expired); it holds itself on until I0,5 gives a signal (this happens when the barrier is down). O0,2 is an inhibiting signal (raise exit barrier).

Label 4 — I0,4 gives a signal when the exit barrier is fully up. This starts on-delay timer T2. After 6 s internal relay B2 is turned on.

Label 5 — C1 is the vehicle counter. The Telemecanique counter gives an output signal at its D (done) terminal only when the target value is reached. The count value is set during programming. The absence of an output from the counter allows O0,4 (SPACE lamp) to be turned on immediately. Each new signal from I0,2 (entry barrier up) causes the counter to count up. When five cars have passed through, the counter signal to O0,5 is turned on and O0,4 is forced off. Each new signal from I0,4 (exit barrier up) counts it down and restores the signal to O0,4.

10.8 Text display

The text display is to cater for up to 32 messages, one of which incorporates two-digit data. In the following description we assume the use of a Brodersen UCT-35P display, which was introduced in Section 3.4.

To transfer up to 32 messages in binary format, we need five outputs ($2^5 = 32$). Those same outputs suffice to generate two-digit data, but we need one more output to set the display in data mode; thus we need six outputs in all. Output signals to a text display are switched very frequently and this represents a difficult *mechanical* duty for relay outputs, hence our choice of transistor outputs.

The messages are easily created using either a keyboard (XT/AT type) or a personal computer running the software supplied with the product. Figure 10.10 shows the appearance of the computer screen and some of the functions available. Within any message, the user can reserve space for up to 16 digits of data (so-called variables) and up to 10 spaces for year, month, day, hour and minute (derived from an internal real-time clock/calendar).

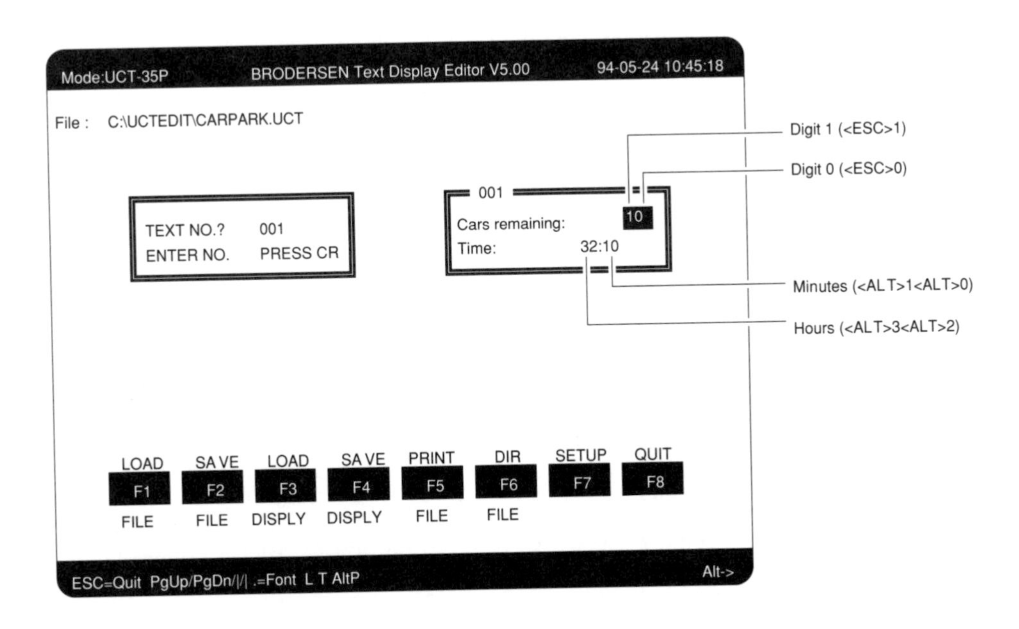

Figure 10.10. Entering the messages and data with a personal computer.

Once the messages are created and stored in the display, the next task is to modify the control program so that the messages can be called up. To do this we need to know what the display expects in the way of output signals from the programmable controller. This information is contained in Fig. 10.11.

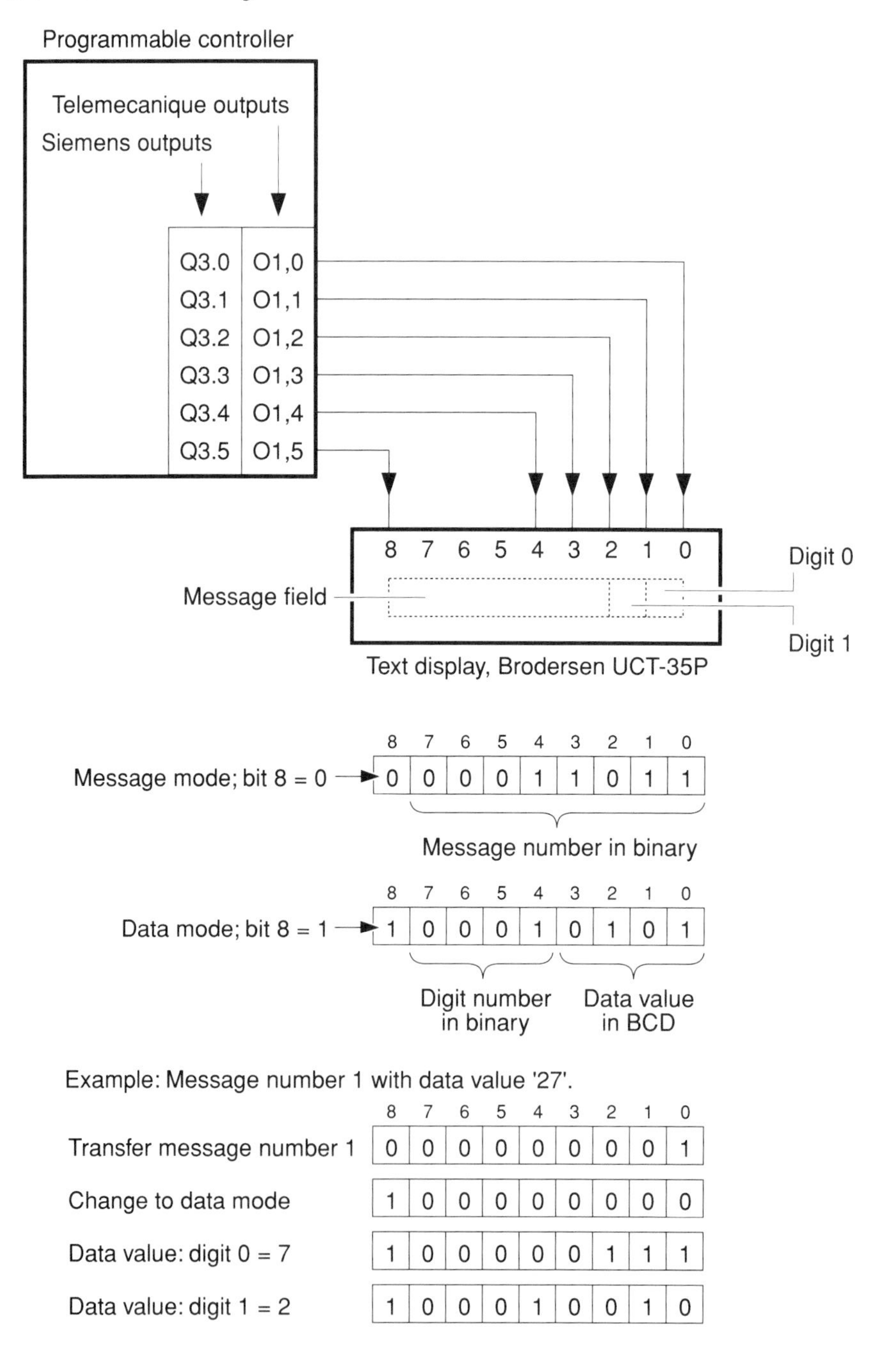

Figure 10.11. How the display interprets the programmable controller's output signals.

The display has two distinct modes, message mode and data mode. The mode is determined by bit 8; 0 (off) = message mode, 1 (on) = data mode. In message mode bits 0–7 (in binary) are interpreted as the message number to be called. In data mode, bits 0–3 are interpreted as the data value to be displayed, while bits 4–7 specify which digit is in question—we need only two.

The example given in Fig. 10.11 shows the transfer of message number 1 with the embedded data '27'. It happens in four stages:

1. in message mode, bits 0–7 yield message number 1 in binary
2. in data mode, bits 0–7 are all 0 to clear the way for new data
3. in data mode, bits 4–7 give the digit number '0' (i.e. the units) and bits 0–3 give the data value '7' in BCD
4. in data mode, bits 4–7 give the digit number '1' (i.e. the tens) and bits 0–3 give the data value '2' in BCD.

Figure 10.12 shows examples of control programs for driving the text display. The programs are offered in one language for each controller because some of the functions are either difficult or impossible to implement in the other language(s).

Figure 10.12(*a*) shows the easiest case, activation of a message without data (number 5). In the Telemecanique case, B135 causes '5' to be transferred to internal word W10 where it is stored in binary form. This in turn is transferred to five outputs beginning at O1,0. An internal word (or data word) stores numbers or other data in the programmable controller's memory. In general, any data being manipulated in the controller must be channelled through a word. In the Siemens case, '5' is unconditionally loaded as a fixed-point number and then transferred to output byte QB3, i.e. the eight outputs Q3.0 to Q3.7.

Figure 10.12(*b*) gives one solution for the car-park project, which requires the transfer of message number 1 with two-digit data. The apparent complication is due to managing the data, but it needs to be programmed only once and can be 'borrowed' for other messages as required. Some of the functions are given without a full explanation since it is beyond the scope of this book, but the main tasks being performed are clearly stated.

10.9 Exercises for Chapter 10 (Solutions in Appendix 1)

10.1. Develop a control program for the Mitsubishi or Sprecher+Schuh controller.

10.2. Describe, in the circuit diagram and in the program implementation, how the loss of pressure in the compressed air system could be catered for.

10.3. Provide for the following additional facilities: at 17.30 each day all inward traffic is stopped and the car-park is prepared for clear-out by these means

- the coin switch is ignored
- the exit barrier is raised constantly
- the car-park sign displays 'FULL' regardless of the number of cars in the car-park.

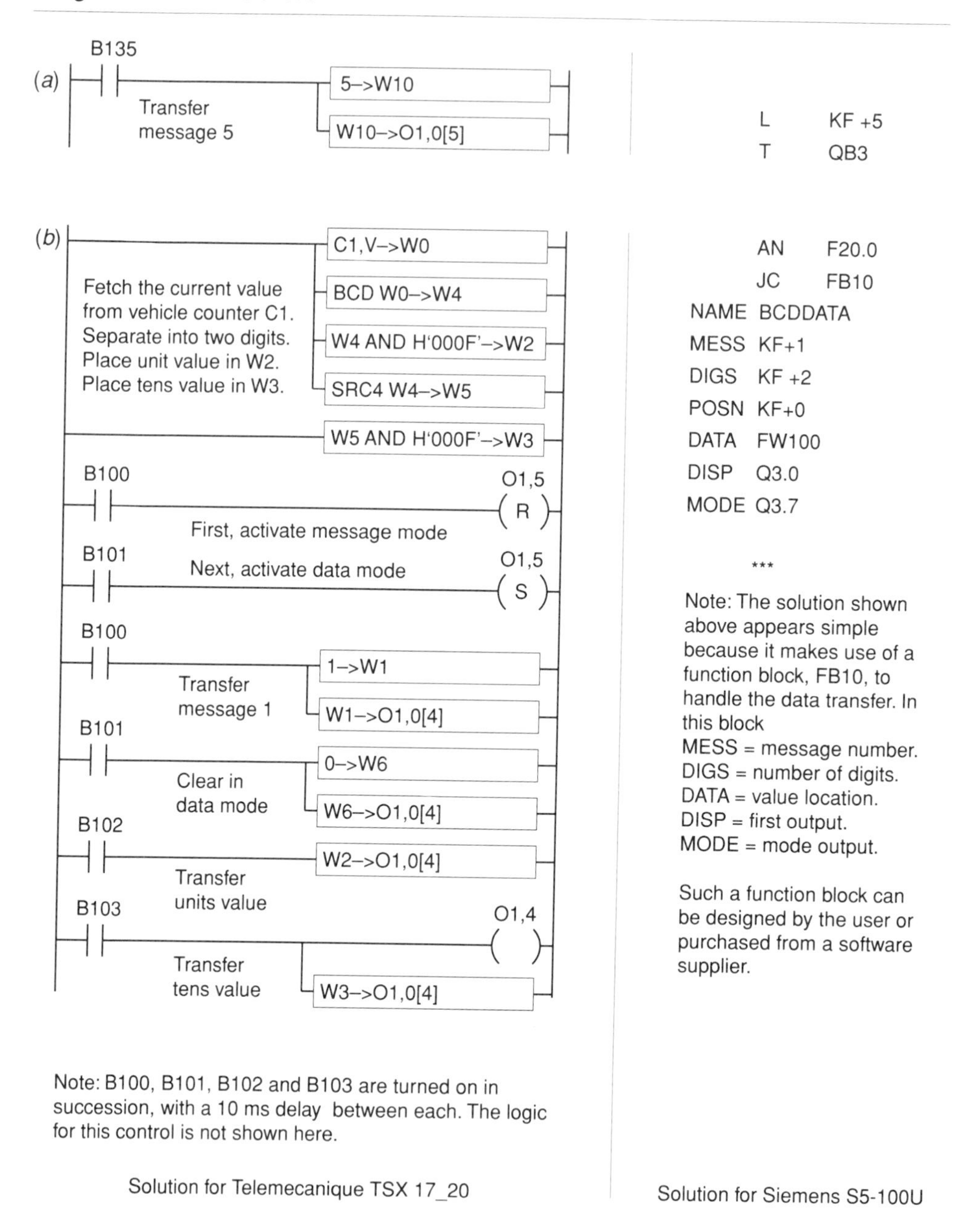

Figure 10.12. Programs for sending (a) a simple message and (b) a message with data.

10.4. The installation offers no protection for the vehicle against a barrier descending at an untimely moment! You are required to:

- propose the necessary sensor/s for detecting such an occurrence
- add the new devices to the I/O list
- incorporate the new device(s) in the control circuit diagram
- provide for the necessary changes in the program (on detecting a vehicle beneath the barrier, the control system must raise the barrier immediately and keep it up until the vehicle has moved away).

10.5. Write out the steps to be taken in implementing the automation of this project. (The implementation topic was omitted from the text to provide for this exercise.)

Chapter 11 Scrap hoist project

11.1 Project description

The installation comprises an electric hoist with a lifting electromagnet which is used for transferring scrap metal from a storage bin to a removal truck. Figure 11.1 shows the physical layout.

The system is intended to function automatically and will fully load the removal truck with six transfers. Three drop positions are identified; two drops are to be made at each of these positions to distribute the load in the truck.

11.2 Equipment

- A programmable controller to automate the process.
- The hoist motor has an 'up' contactor K1 and a 'down' contactor K2.
- The traverse motor has a 'left' contactor K3 and a 'right' contactor K4.
- The electromagnet is energized by contactor K5.
- A beacon H1 is provided to signal the truck to depart with a full load.

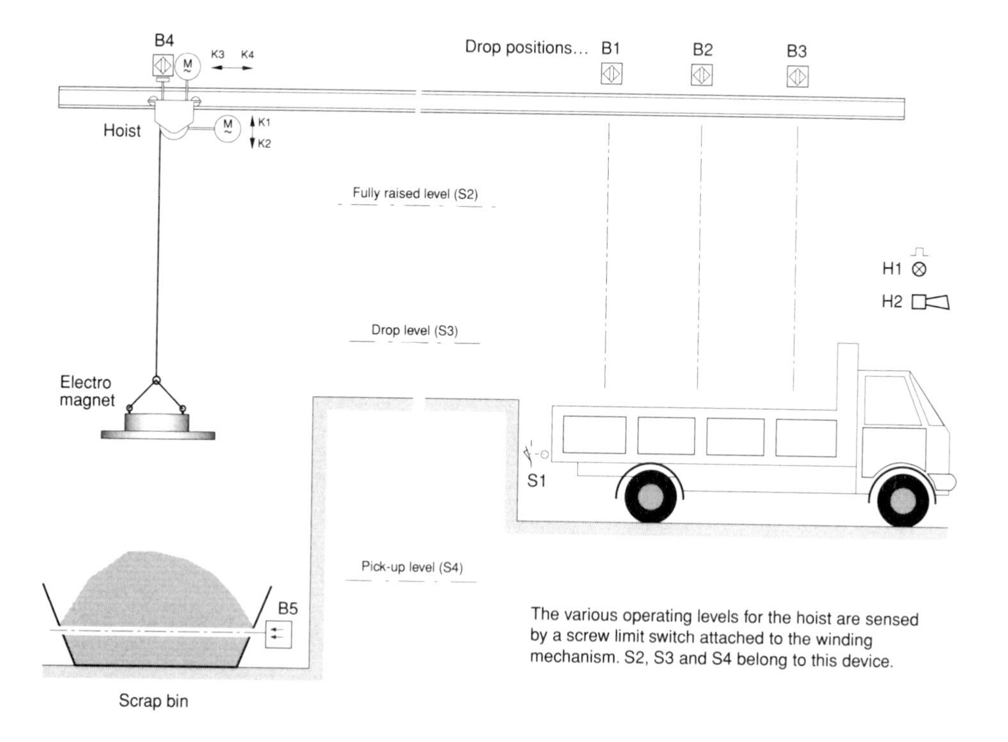

Figure 11.1. Scrap hoist installation.

- An alarm sounder H2 is provided to signal certain fault conditions.
- Drop positions are identified by proximity sensors B1, B2 and B3.
- The 'home' position is identified by proximity sensor B4.
- Photoelectric sensor B5 is used to verify that scrap is present.
- Limit switch S1 is closed when the truck is in place.
- The hoist is fitted with a screw limit switch having three separate contacts:
    1. S2 corresponding to the 'fully up' level;
    2. S3 corresponding to the 'drop' level;
    3. S4 corresponding to the 'pick-up' level.

11.3 Specification

1. Control is to be automatic. The cycle begins with the hoist lifting fully up and then moving to the home position.
2. With scrap available and the truck in place, the hoist makes its first pick-up, lifts fully and then moves to the first drop position, lowering its load to the drop level before discharging.
3. It must lift fully and return for the next pick-up, which in due course will be discharged at the second drop position.
4. Lifting fully again, it returns for the third pick-up which is destined to be discharged at the third drop position.
5. Steps 2, 3 and 4 are then repeated to build a full load for the truck.
6. After the final drop the hoist lifts fully and returns home. The beacon then flashes and continues to do so until the truck moves away with its load. A new cycle begins when the truck returns.
7. The alarm sounder operates under either of the following conditions:
    - an overload occurs on the hoist motor
    - an overload occurs on the traverse motor.

11.4 Solution

We consider three aspects:

- the main circuit, including switchgear and protection for the motors and the electromagnet as well as power supplies
- the control circuit, including I/O connections, which is specific to the controller used
- the control program, which is specific to the controller used.

11.5 Main circuit

The main circuit is described in Fig. 11.2. The 115 V a.c. supply serving the contactor coils, the beacon, the sounder and the programmable controller is derived from a control circuit transformer (not shown). Safety devices are omitted for simplicity and clarity.

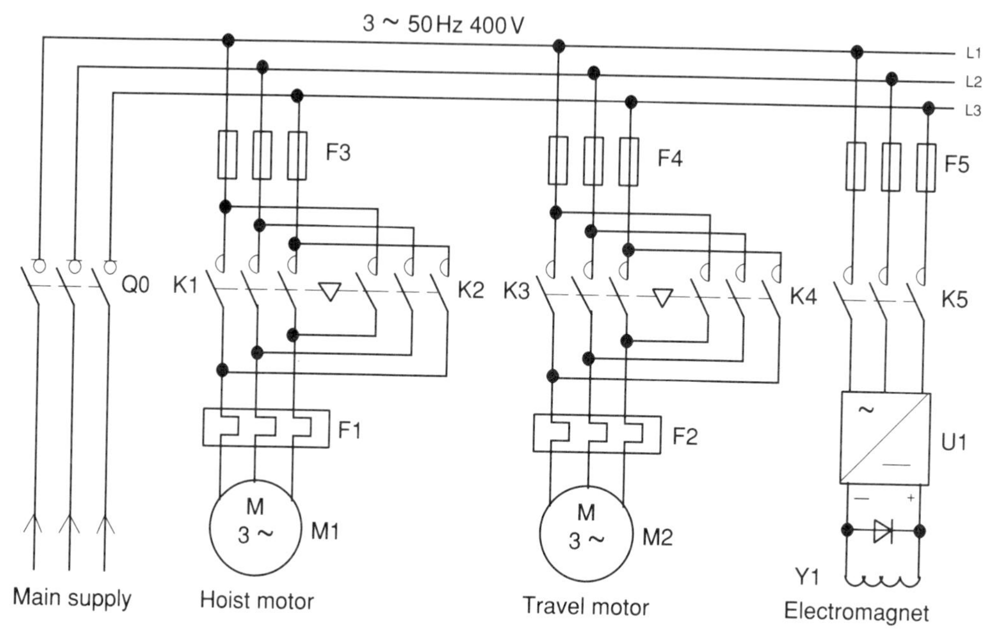

Figure 11.2. Main circuit.

11.6 Control circuit

Tables 11.1 and 11.2 give the I/O allocation for all four controllers. Control circuits for two of them are shown in Figs 11.3 and 11.4. The Mitsubishi controller featured is the Fx-48MR. This model has nearly double the I/O capacity of its predecessor (in the F_2 range) for a given case size and possesses superior programming, calculation, communication and diagnostic functions.

The inputs are 24 V d.c. Each proximity sensor is a three-wire PNP type, giving an input signal when the target is nearby. Photo-sensor B5 is similar and gives an input signal when scrap is present. Each contact of the screw limit switch gives a signal (of adequate duration) when the hoist is at the relevant level.

11.7 Program

Solutions are offered for the Mitsubishi and Sprecher+Schuh controllers. For this project we have included the explanations with the illustrations. By sectionalizing the project in this way, we hope to facilitate a better understanding of the program.

An important function in this project is control of the loading *sequence*. There are several possibilities for doing this, and we have chosen to use a different approach for each controller to demonstrate two of them. They give equally satisfactory results. A third possibility would be a sequential function chart (SFC). Because this is a convenient way of

Table 11.1. Schedule of inputs

Device	Item	Input			
		Mitsubishi	Siemens	Sprecher+Schuh	Telemecanique
Drop position 1	B1	X1	I0.1	X001	I0,1
Drop position 2	B2	X2	I0.2	X002	I0,2
Drop position 3	B3	X3	I0.3	X003	I0,3
Home position	B4	X4	I0.4	X004	I0,4
Scrap present	B5	X5	I0.5	X005	I0,5
Truck in place	S1	X6	I0.6	X006	I0,6
Hoist up level	S2	X7	I0.7	X007	I0,7
Hoist drop level	S3	X10	I1.0	X008	I0,8
Hoist pick level	S4	X11	I1.1	X009	I0,9
Hoist overload	F1	X12	I1.2	X010	I0,10
Travel overload	F2	X13	I1.3	X011	I0,11

Table 11.2. Schedule of outputs

Device	Item	Output			
		Mitsubishi	Siemens	Sprecher+Schuh	Telemecanique
Raise hoist	K1	Y0	Q2.0	Y001	O0,0
Lower hoist	K2	Y1	Q2.1	Y002	O0,1
Travel left	K3	Y2	Q2.2	Y003	O0,2
Travel right	K4	Y3	Q2.3	Y004	O0,3
Electromagnet	K5	Y4	Q2.4	Y005	O0,4
'Truck loaded'	H1	Y5	Q2.5	Y006	O0,5
Alarm	H2	Y6	Q2.6	Y007	O0,6

explaining the requirements of the system it is shown in Fig. 11.5; for controllers equipped with SFC programming it could also be the basis of the control program. Function chart programming is explained in Chapter 6. It may be helpful to recall that:

- on switch-on, the step with the double box (step 0 here) is active by default
- when a step is active, the associated actions are carried out
- the only transition that matters is the one immediately *after* the active step.

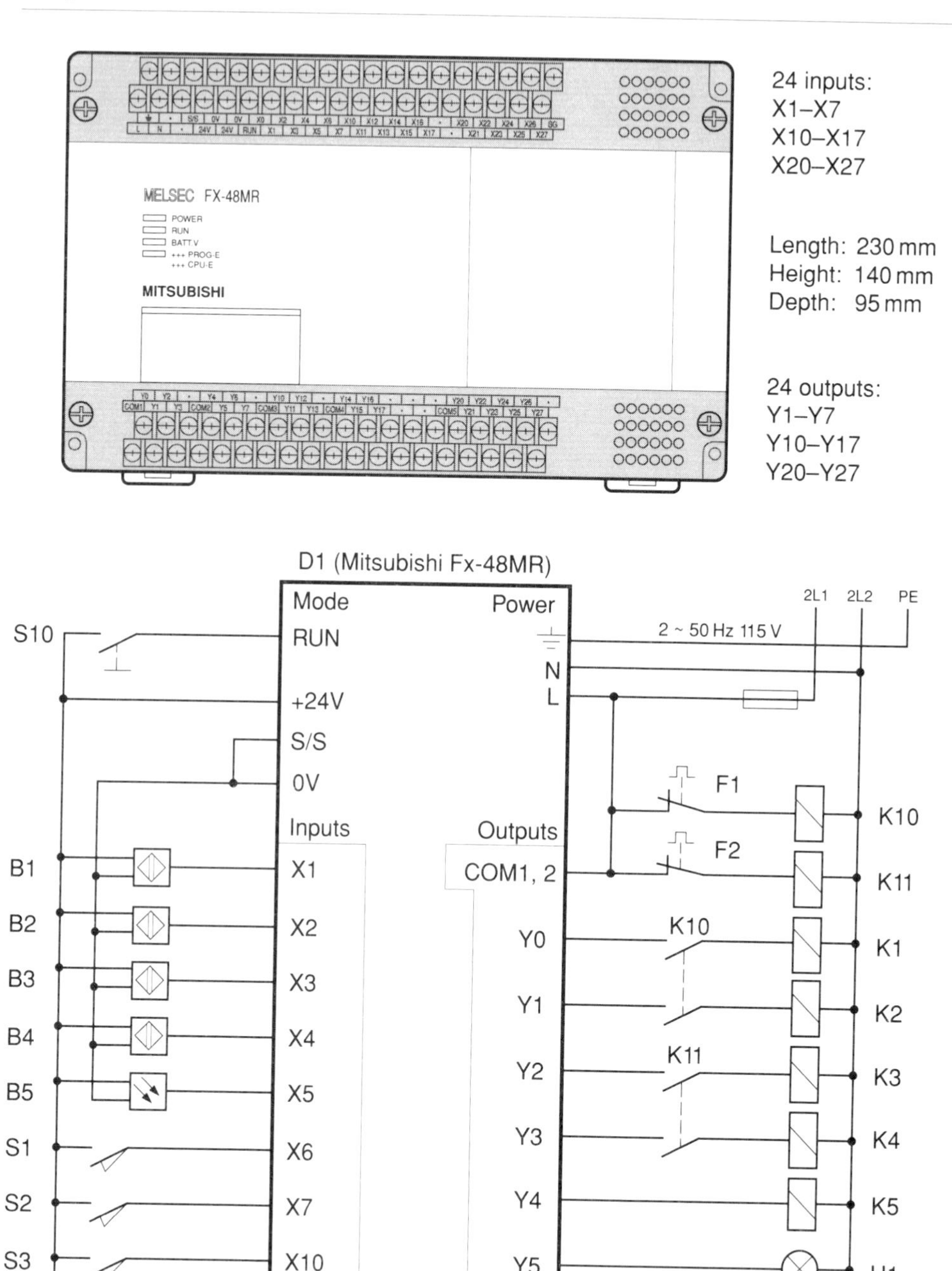

Figure 11.3. Arrangement and connection of the Mitsubishi Fx-48 controller.

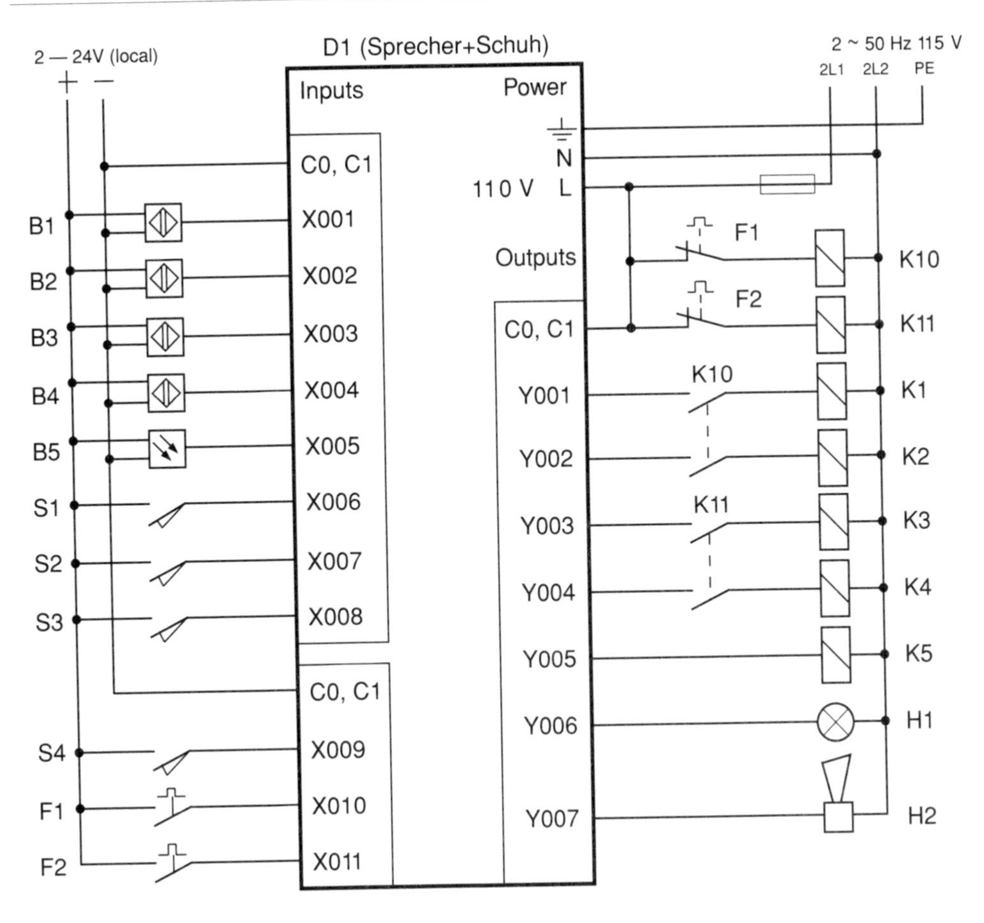

Figure 11.4. Connection of the Sprecher+Schuh SESTEP® 290 controller.

The proposed solution for the Mitsubishi Fx-48MR is given in ladder language in Figs 11.6 to 11.10.

The overall program is divided into nine sections each dealing with a specific task or function. Sequence control is based on a shift register. A *register* is part of the controller's memory normally reserved for storing a value (data). We have used one, together with register-specific instructions, to manipulate signals in such a way that it mimics the required control sequence. Figure 11.8 explains this.

The proposed solution for the Sprecher+Schuh SESTEP® 290 is given in ladder language in Figs 11.11 to 11.14.

The overall program is divided into nine sections, each dealing with a specific task or function. Sequence control is based on a counter (CNT V3) with which special elements S300, S301, S302 … are associated. Our strategy is, in effect, to count the number of drops made (six are required) and to let the counter dictate the actions appropriate to that number.

We do not give exercises in this chapter in view of the complexity of the program.

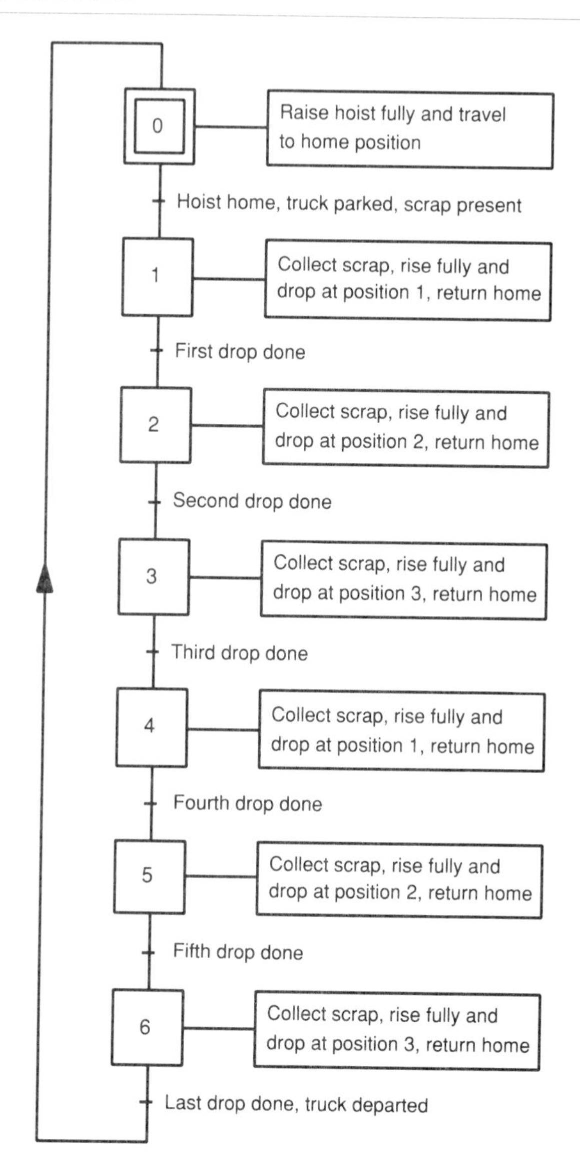

Figure 11.5. Sequential function chart for the loading cycle.

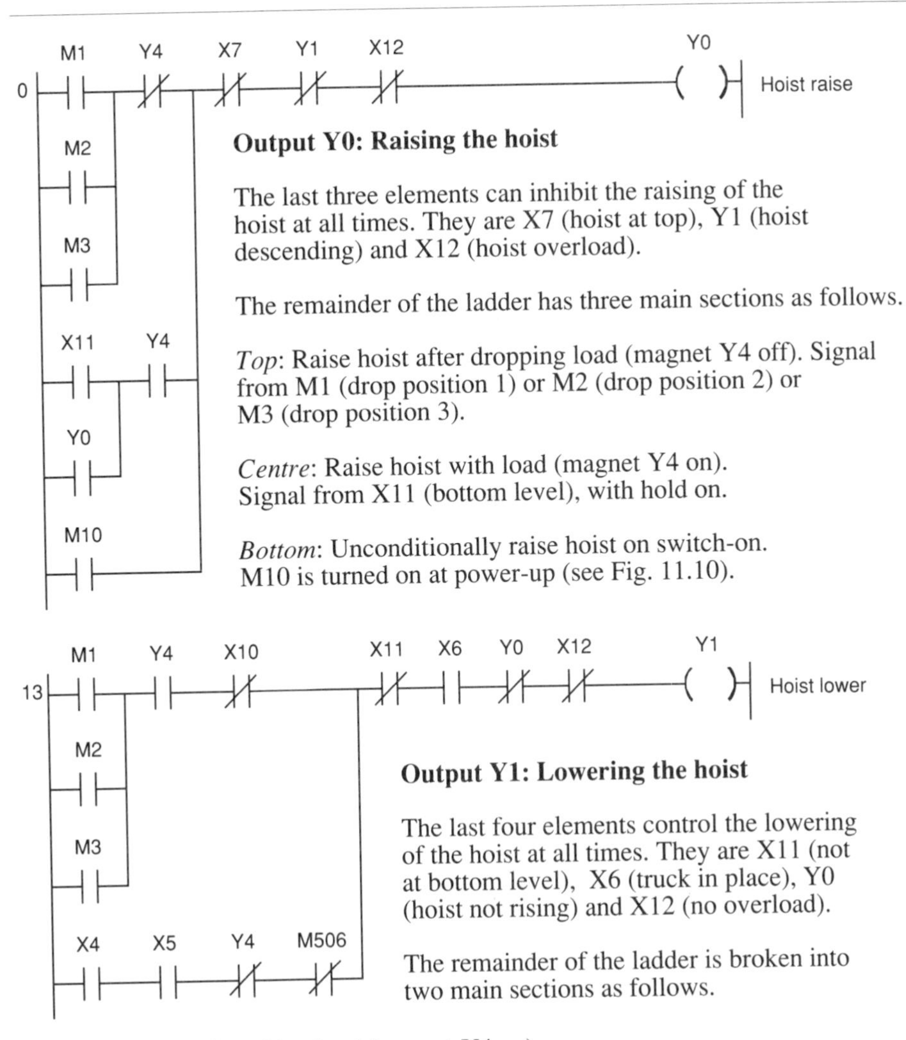

Output Y0: Raising the hoist

The last three elements can inhibit the raising of the hoist at all times. They are X7 (hoist at top), Y1 (hoist descending) and X12 (hoist overload).

The remainder of the ladder has three main sections as follows.

Top: Raise hoist after dropping load (magnet Y4 off). Signal from M1 (drop position 1) or M2 (drop position 2) or M3 (drop position 3).

Centre: Raise hoist with load (magnet Y4 on). Signal from X11 (bottom level), with hold on.

Bottom: Unconditionally raise hoist on switch-on. M10 is turned on at power-up (see Fig. 11.10).

Output Y1: Lowering the hoist

The last four elements control the lowering of the hoist at all times. They are X11 (not at bottom level), X6 (truck in place), Y0 (hoist not rising) and X12 (no overload).

The remainder of the ladder is broken into two main sections as follows.

Top: Lower hoist with a load (magnet Y4 on).
A signal from M1 (drop position 1) or M2 (drop position 2) or M3 (drop position 3) and not X10 (drop level).

Bottom: Lower hoist for picking up a load (magnet Y4 off). Signals from X4 (home position) and X5 (scrap present) but no signals from M506 (M506 is on when the truck is full).

Figure 11.6. Ladder solution for the Mitsubishi controller.

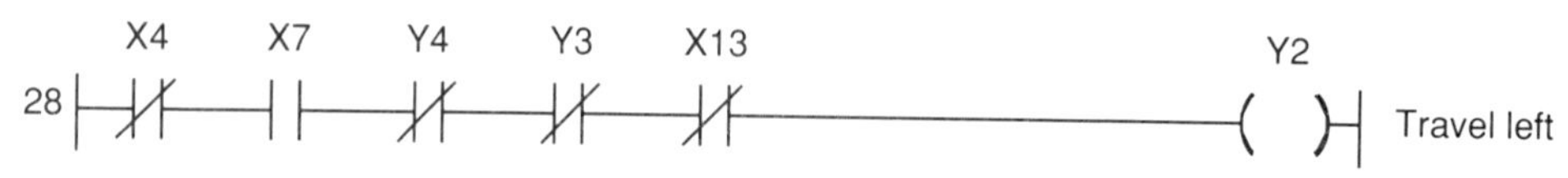

Output Y2: Crane travel left

This is possible when all five of these conditions are met: No signal from X4 (home position); signal from X7 (the magnet must be at top level); no signal from Y4 (the magnet must be off); no signal from Y3 (crane moving right) and no signal from X13 (overload relay trip).

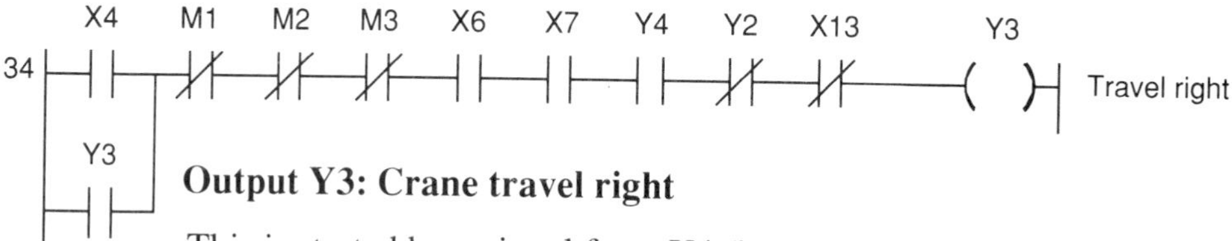

Output Y3: Crane travel right

This is started by a signal from X4 (home position) and is held on. The enabling signals come from X6 (truck in place), X7 (magnet at top level) and Y4 (magnet on). Inhibiting signals are Y2 (travel left) and X13 (overload trip). Normally, one of the signals M1 (drop position 1), M2 (drop position 2) or M3 (drop position 3) will act as the stop signal.

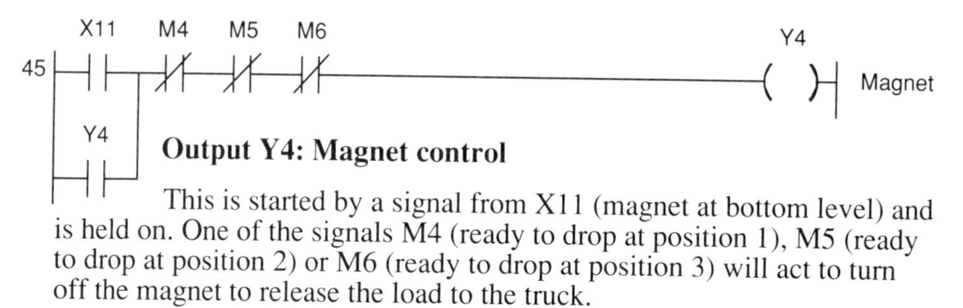

Output Y4: Magnet control

This is started by a signal from X11 (magnet at bottom level) and is held on. One of the signals M4 (ready to drop at position 1), M5 (ready to drop at position 2) or M6 (ready to drop at position 3) will act to turn off the magnet to release the load to the truck.

Figure 11.7. Ladder solution for the Mitsubishi controller (continued).

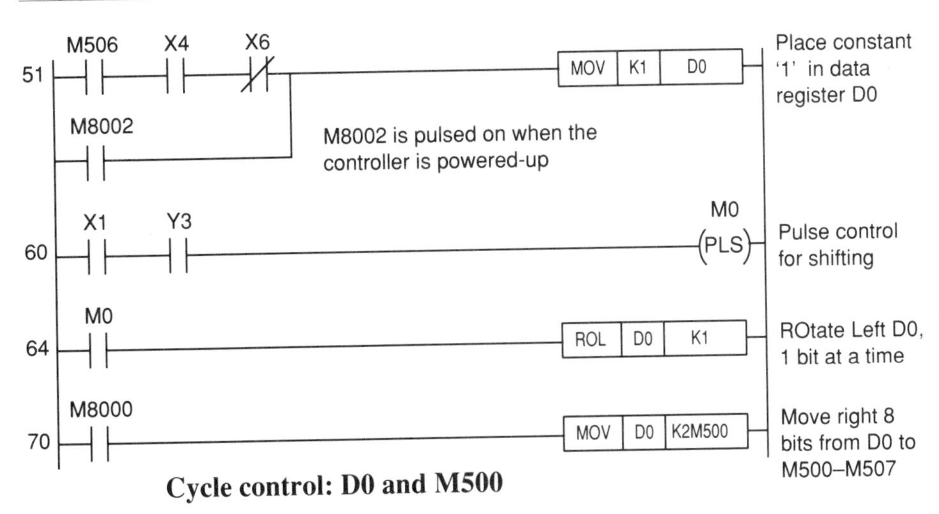

Cycle control: D0 and M500

A shift register is used for cycle control. A register stores 16 signals (bits) which may be on (= 1) or off (= 0). Collectively these bits are called a *data word* and the Fx-48 can store 512 such words in registers D0–D511.

In the top rung M8002 (pulsed on power-up) causes '1' to be moved to D0, giving the initial conditions shown below. An eventual return to these conditions is made by M506 (cycle done), X4 (home position), and not X6 (truck departed).

The next two rungs specify the conditions for shifting D0. Every time X1 (crane passing B1) and Y3 (moving right) give a signal, M0 is pulsed, activating ROL (rotate left) on D0. With each pass of the hoist the '1' is shifted one place to the left.

Finally the bottom rung transfers the right-hand eight bits to the internal relays M500–M507, which are directly accessible for control purposes.

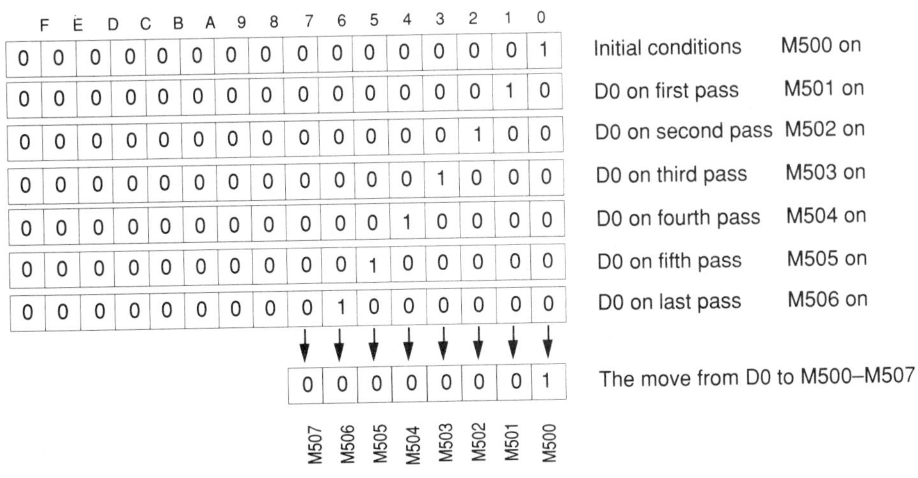

F	E	D	C	B	A	9	8	7	6	5	4	3	2	1	0		
0	0	0	0	0	0	0	0	0	0	0	0	0	0	0	1	Initial conditions	M500 on
0	0	0	0	0	0	0	0	0	0	0	0	0	0	1	0	D0 on first pass	M501 on
0	0	0	0	0	0	0	0	0	0	0	0	0	1	0	0	D0 on second pass	M502 on
0	0	0	0	0	0	0	0	0	0	0	0	1	0	0	0	D0 on third pass	M503 on
0	0	0	0	0	0	0	0	0	0	0	1	0	0	0	0	D0 on fourth pass	M504 on
0	0	0	0	0	0	0	0	0	0	1	0	0	0	0	0	D0 on fifth pass	M505 on
0	0	0	0	0	0	0	0	0	1	0	0	0	0	0	0	D0 on last pass	M506 on

M507	M506	M505	M504	M503	M502	M501	M500	
0	0	0	0	0	0	0	1	The move from D0 to M500–M507

Figure 11.8. Ladder solution for the Mitsubishi controller (continued).

76 M501 X1 M1 Drop position 1
M504

80 M502 X2 M2 Drop position 2
M505

84 M503 X3 M3 Drop position 3
M506

Establishing the drop positions

Elements M501–M506 belong to the shift register described above and serve to indicate on which pass the crane is currently running. Thus M1 is turned on when the count is 1 or 4 (the first or fourth pass of the crane) and there is a signal from X1 (position 1). M2 is turned on when the count is 2 or 5 (the second or fifth pass of the crane) and there is a signal from X2 (position 2). M3 is turned on when the count is 3 or 6 (the third or sixth pass of the crane) and there is a signal from X3 (position 3).

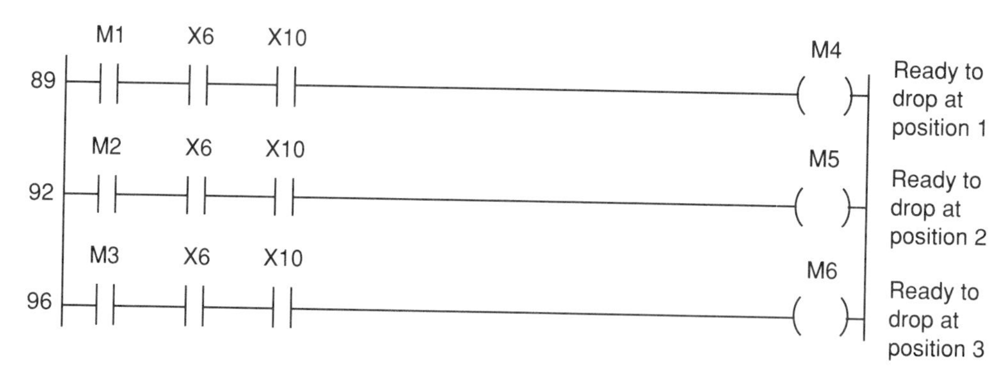

Ready to drop load at position

The signal 'ready to drop load' is established for each of the three positions. M1, M2 and M3 are the drop position signals developed above. Input X6 is included to ensure the truck is in place and X10 to ensure the hoist has descended to the correct level for dropping the load.

Figure 11.9. Ladder solution for the Mitsubishi controller (continued).

M506 Y4 X4 T0 Y5
100
Full load indicator

Output Y5: Full load beacon

This is turned on with a signal from M506 (loading cycle complete), no signal from Y4 (magnet must be off) and a signal from X4 (home position). T0 belongs to the flasher developed below.

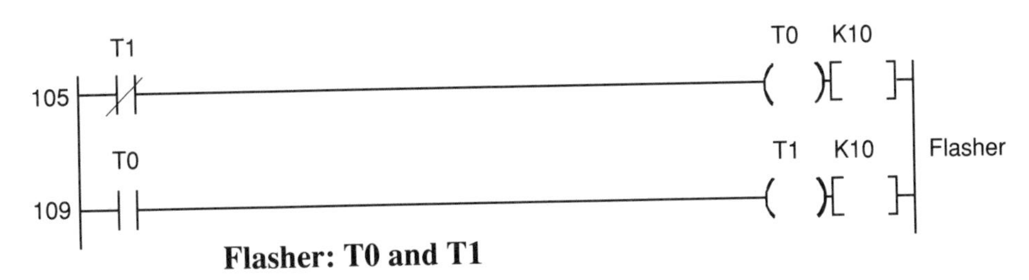

Flasher: T0 and T1

The flasher is needed for the full load beacon. We show here how two timers can be assembled to create a flasher. T0 is first to be started; 1s later it times out and starts T1. One second later T1 times out and cuts off T0 which resets the whole timing circuit. Repeating occurs indefinitely.

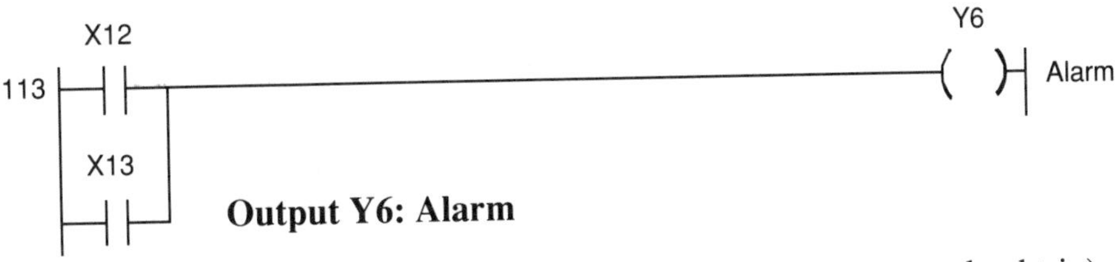

Output Y6: Alarm

The alarm is turned on by a signal from X12 (hoist motor overload trip) or X13 (travel motor overload trip).

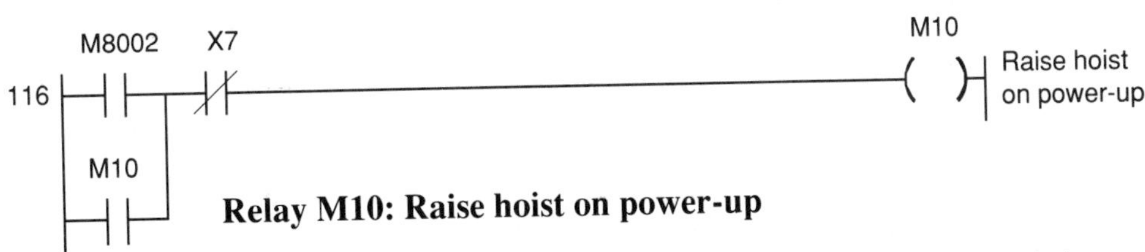

Relay M10: Raise hoist on power-up

M10 is turned on by M8002 (pulsed on when the controller powers up); it then holds itself on until a signal from X7 (hoist at top level) appears.

Figure 11.10. Ladder solution for the Mitsubishi controller (continued).

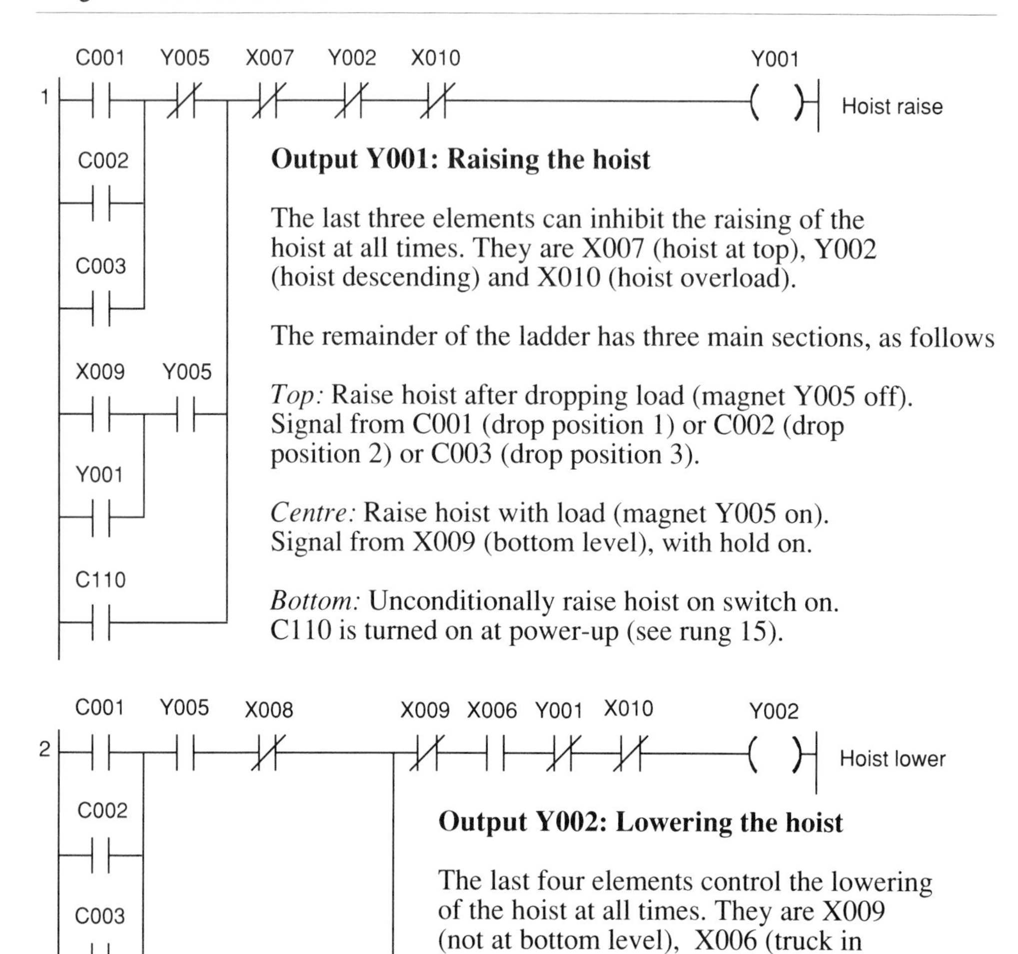

Output Y001: Raising the hoist

The last three elements can inhibit the raising of the hoist at all times. They are X007 (hoist at top), Y002 (hoist descending) and X010 (hoist overload).

The remainder of the ladder has three main sections, as follows

Top: Raise hoist after dropping load (magnet Y005 off). Signal from C001 (drop position 1) or C002 (drop position 2) or C003 (drop position 3).

Centre: Raise hoist with load (magnet Y005 on). Signal from X009 (bottom level), with hold on.

Bottom: Unconditionally raise hoist on switch on. C110 is turned on at power-up (see rung 15).

Output Y002: Lowering the hoist

The last four elements control the lowering of the hoist at all times. They are X009 (not at bottom level), X006 (truck in place), Y001 (hoist not rising) and X010 (no overload).

The remainder of the ladder is broken into two main sections, as follows

Top: Lower hoist with a load (magnet Y005 on).
A signal from C001 (drop position 1) or C002 (drop position 2) or C003 (drop position 3) and not X008 (drop level).

Bottom: Lower hoist for picking up a load (magnet Y005 off).
Signals from X004 (home position) and X005 (scrap present) but no signals from C100 (C100 is on when the truck is full).

Figure 11.11. Ladder solution for the Sprecher+Schuh controller.

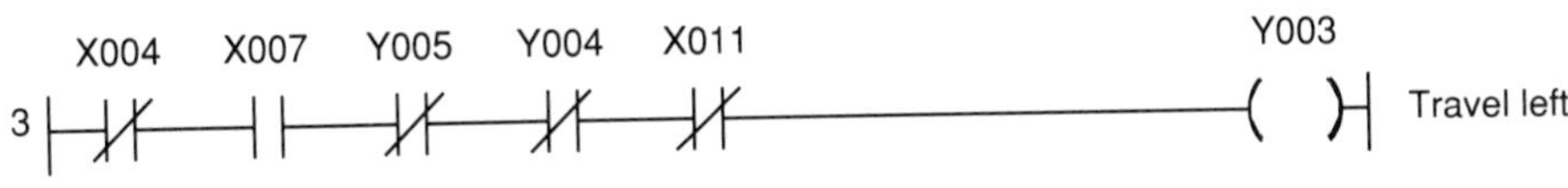

Output Y003: Crane travel left

This is possible when all five of these conditions are met:
No signal from X004 (home position); signal from X007 (the magnet must be at top level); no signal from Y005 (the magnet must be off); no signal from Y004 (crane moving right) and no signal from X011 (overload relay trip).

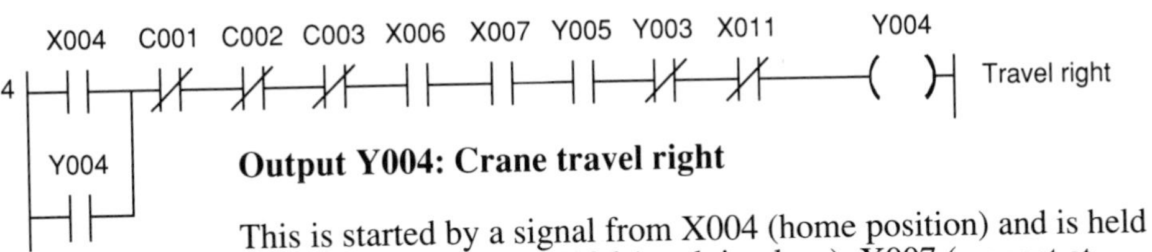

Output Y004: Crane travel right

This is started by a signal from X004 (home position) and is held on. The enabling signals come from X006 (truck in place), X007 (magnet at top level) and Y005 (magnet on). Inhibiting signals are Y003 (travel left) and X011 (overload trip). Normally, one of the signals C001 (drop position 1), C002 (drop position 2) or C003 (drop position 3) will act as the stop signal.

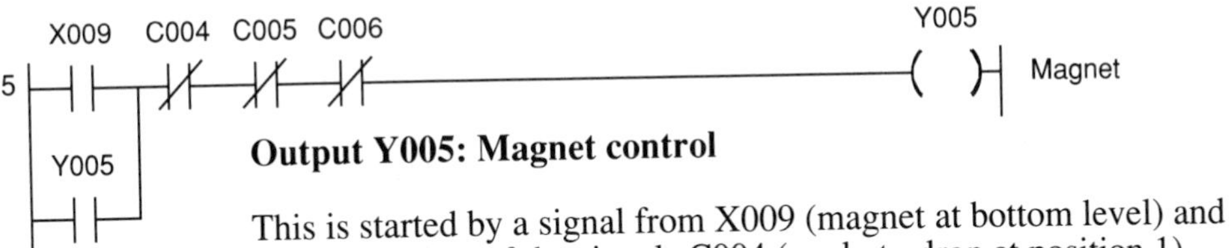

Output Y005: Magnet control

This is started by a signal from X009 (magnet at bottom level) and is held on. One of the signals C004 (ready to drop at position 1), C005 (ready to drop at position 2) or C006 (ready to drop at position 3) will act to turn off the magnet to release the load to the truck.

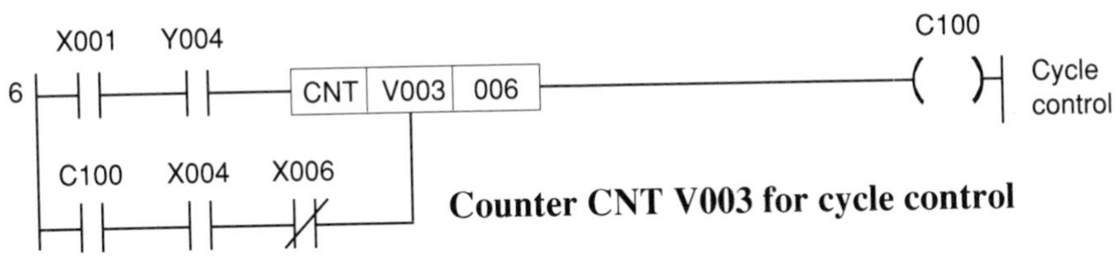

Counter CNT V003 for cycle control

This counter has a preset value of 6 to record the six passes the crane makes in one cycle. Signals from X001 (position 1) and Y004 (travelling right) are used to count up. When the count reaches 6 relay C100 is turned on to record 'cycle done'. Reset occurs when the cycle is done, the hoist is in the home position and the truck moves away.

Figure 11.12. Ladder solution for the Sprecher+Schuh controller (continued).

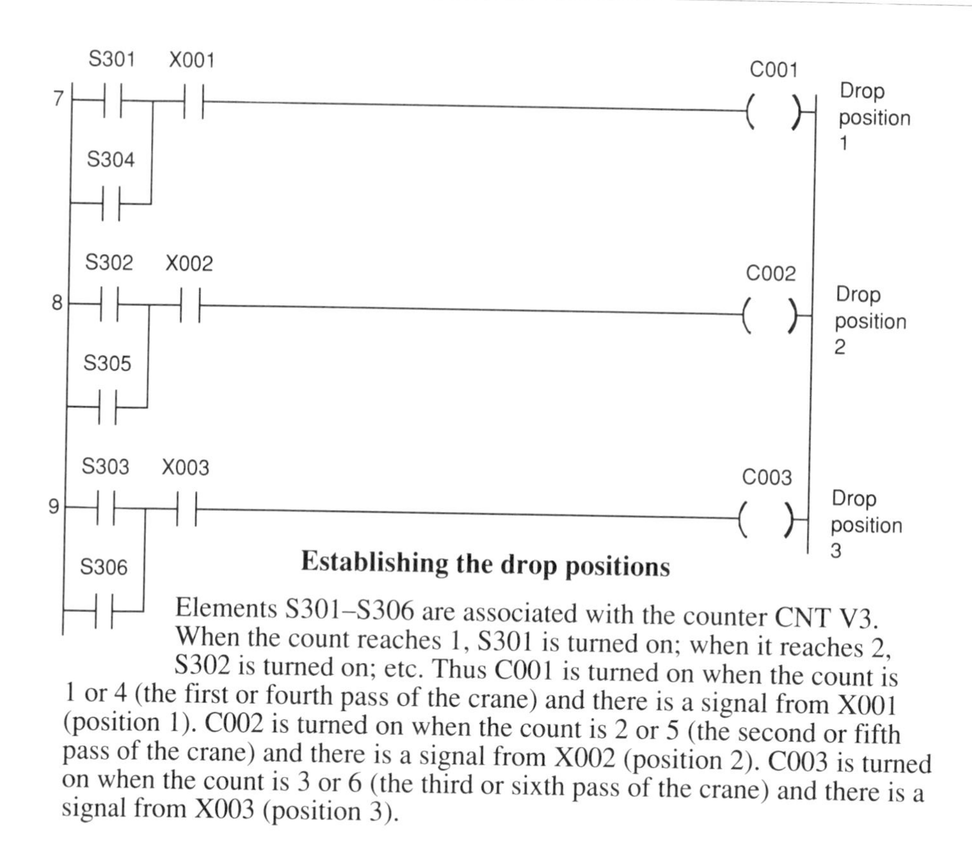

Establishing the drop positions

Elements S301–S306 are associated with the counter CNT V3. When the count reaches 1, S301 is turned on; when it reaches 2, S302 is turned on; etc. Thus C001 is turned on when the count is 1 or 4 (the first or fourth pass of the crane) and there is a signal from X001 (position 1). C002 is turned on when the count is 2 or 5 (the second or fifth pass of the crane) and there is a signal from X002 (position 2). C003 is turned on when the count is 3 or 6 (the third or sixth pass of the crane) and there is a signal from X003 (position 3).

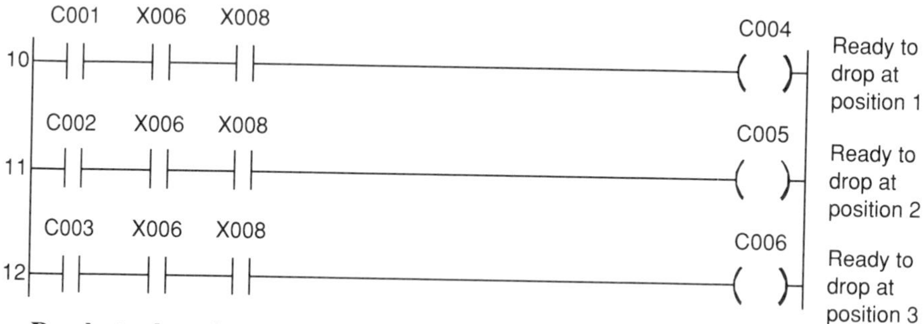

Ready to drop load at position...

The signal 'ready to drop load' is established for each of the three positions. C001, C002 and C003 are the drop position signals developed above. Input X006 is included to ensure the truck is in place and X008 to ensure the hoist has descended to the correct level for dropping the load.

Figure 11.13. Ladder solution for the Sprecher+Schuh controller (continued).

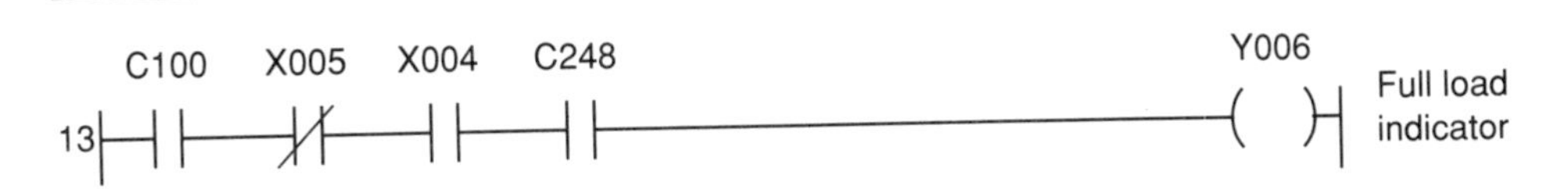

Output Y006: Full load beacon

This is turned on with a signal from C100 (loading cycle complete), no signal from X005 (magnet must be off) and a signal from X004 (crane back at home position). The element C248 is an internal relay which flashes on/off at 0.8 s intervals. It can be used anywhere in the program and it saves having to build a flasher with two timers.

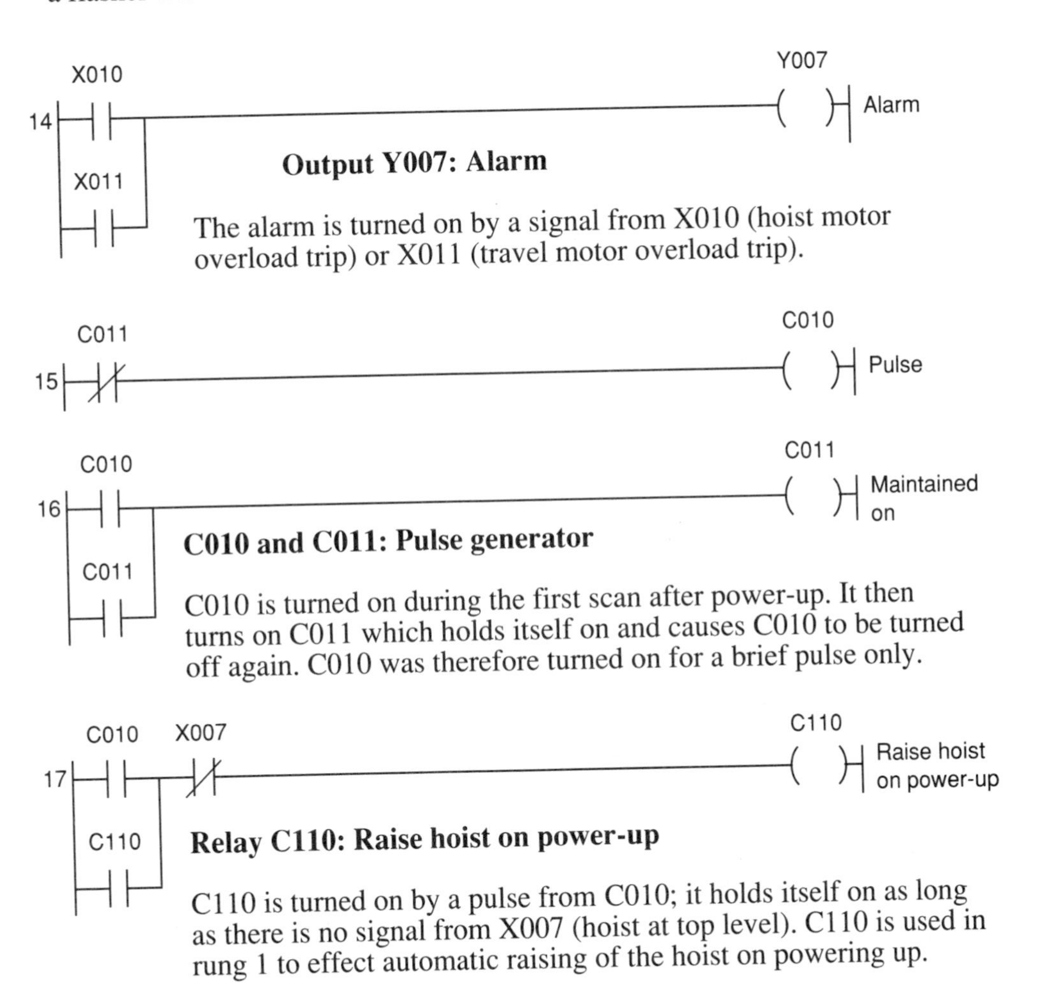

Output Y007: Alarm

The alarm is turned on by a signal from X010 (hoist motor overload trip) or X011 (travel motor overload trip).

C010 and C011: Pulse generator

C010 is turned on during the first scan after power-up. It then turns on C011 which holds itself on and causes C010 to be turned off again. C010 was therefore turned on for a brief pulse only.

Relay C110: Raise hoist on power-up

C110 is turned on by a pulse from C010; it holds itself on as long as there is no signal from X007 (hoist at top level). C110 is used in rung 1 to effect automatic raising of the hoist on powering up.

Figure 11.14. Ladder solution for the Sprecher+Schuh controller (continued).

Chapter 12 Documentation and troubleshooting

12.1 Documentation

One of the strongest arguments for purchasing a graphic programming terminal or a personal computer and software is that it facilitates the printing of the program and related data on a suitable printer (Fig. 12.1). This has several advantages:

- the documentation can be automatically updated
- the complete program can be reviewed
- multiple copies can be provided quickly and easily
- fault-finding can be approached with confidence.

Depending on the sophistication of the terminal, some or all of the following may be capable of being printed:

1. ladder and/or list program (part or all)
2. sequential function chart (where applicable)
3. I/O list with comments
4. cross-reference for selected elements
5. circuit diagram of modules, inputs and outputs
6. a list of configured timers, counters and other elements with their constants or set-up values.

The user also has the option of setting up a title page, the page numbering and the header or footer on each page.

The quality of the printout depends on the type of printer used. A laser printer gives an excellent, sharp copy; a dot matrix printer will try to 'draw' ladder diagrams and charts by using ordinary text shapes (characters) arranged in a favourable way. Figure 12.2 shows the result.

Horizontal lines are printed as	– – + – – + – – + – –
Vertical lines are printed as	! (placed one above another)
Signal tests are printed as	-] [- or -]/[-
Outputs are printed as	-( )-

Clearly, the limitations of the standard printer have at least partly determined the type of symbol with which the ladder diagram has evolved.

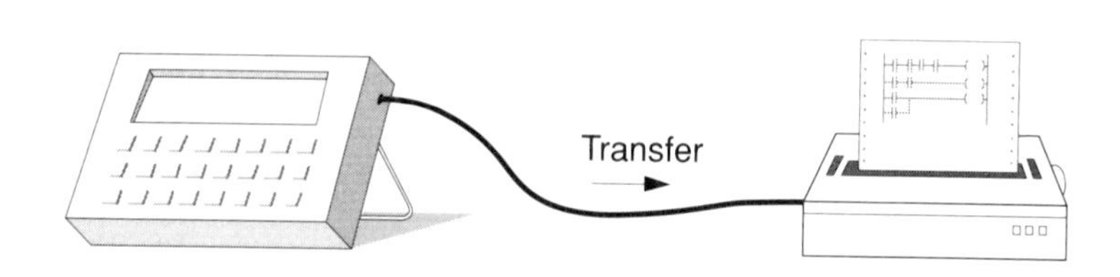

Figure 12.1. The graphic programming terminal can transfer the program to a printer.

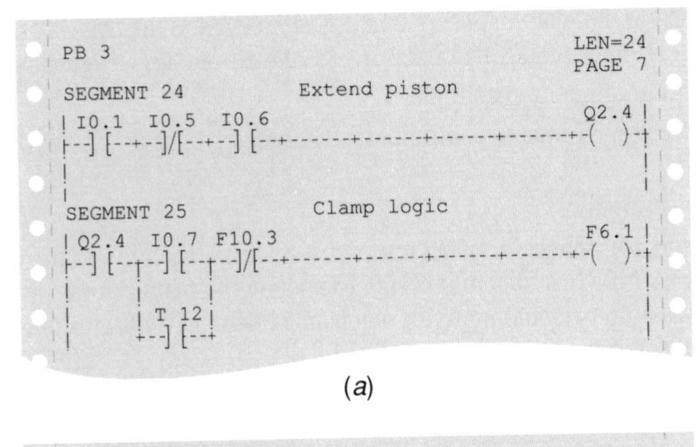

(*a*)

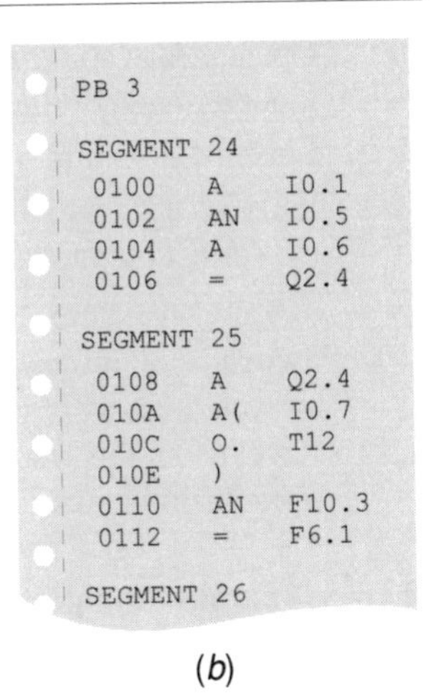

(*b*)

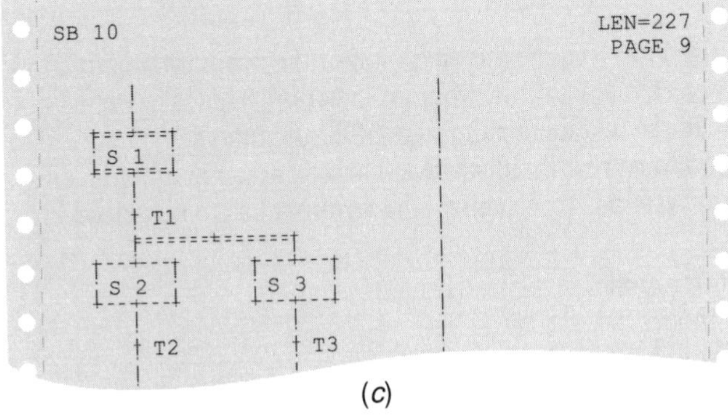

(*c*)

Samples of printout with the Siemens controller:
(*a*) ladder diagram
(*b*) instruction list
(*c*) function chart

Figure 12.2. Samples of printout.

12.2 Pre-commissioning

The commissioning of any project involves carrying out a number of tests in a logical order to ensure that the operation of the plant is as described in the specification. When the plant is controlled by a programmable controller, the following pre-commissioning tests should be carried out.

1. Visually check all plant wiring to ensure that the installation is complete, safe, and to the standards set by the specification and local regulations.
2. With the aid of test instruments check all control wiring to the programmable controller to ensure that it is complete, safe, and connected to the appropriate supply sources. (NB: Do not use a high-voltage test instrument such as a 'Megger', as it can damage the controller and sensors.)
3. Set all protective devices to their appropriate trip settings, including motor protection relays and circuit-breakers.
4. Switch off all motor isolators (disconnectors).
5. Switch on the power supply to the programmable controller.

6. Operate each control switch in the plant to ensure that it is connected to the correct input terminal of the programmable controller and provides the correct logic signal.
7. Load the program to the programmable controller.
8. Force on each output of the programmable controller to check that it is connected to the correct output device.
9. Test all emergency stop push-buttons.
10. Switch on all motor isolators (disconnectors) and repeat test 8; check each motor for correct direction of rotation. (NB: Ensure that it is safe to operate the plant prior to carrying out this test, and that no personal injury or mechanical damage could result from incorrect direction of rotation of motors, etc.)

12.3 Visualizing operations

A programming terminal can be used to visualize the operation of the program, i.e. to allow the user to observe selected parts or elements of the program in action. This is extremely useful for fault-finding: the user can see at a glance the state of all the relevant signals, enabling the cause of a malfunction to be readily identified. There is no comparable facility in electromechanical or pneumatic systems. To visualize, the terminal has to be placed in the correct mode:

- Mitsubishi — monitor mode
- Siemens — test mode
- Sprecher+Schuh — read mode/monitor mode
- Telemecanique — debug mode

Figure 12.3 shows how the Sprecher+Schuh PRG-20 list type terminal is used for visualization. The terminal can call up any step (address) in the program and indicate the status of the signal (on or off) or the value of any parameter (time, count, etc.).

Figure 12.4 shows a graphic type programming terminal in action. This is far more comprehensive than the list terminal; the relationship between the various elements is very obvious and the active elements are highlighted by colour or contrast. In addition the current values of timers and counters and other parameters may be displayed.

Figure 12.5 shows how a sequential function chart is visualized (on the screen of a personal computer running Telemecanique PL7-2 software). The active step (step 3) is in inverse video. If the cycle is not progressing it is easy to find the cause: the transition immediately after the active step holds the key. By zooming in on this transition we see that the signal from internal relay B181 is absent. We must therefore investigate further to find the reason for this. In demanding situations we can force on B181 to clear the transition.

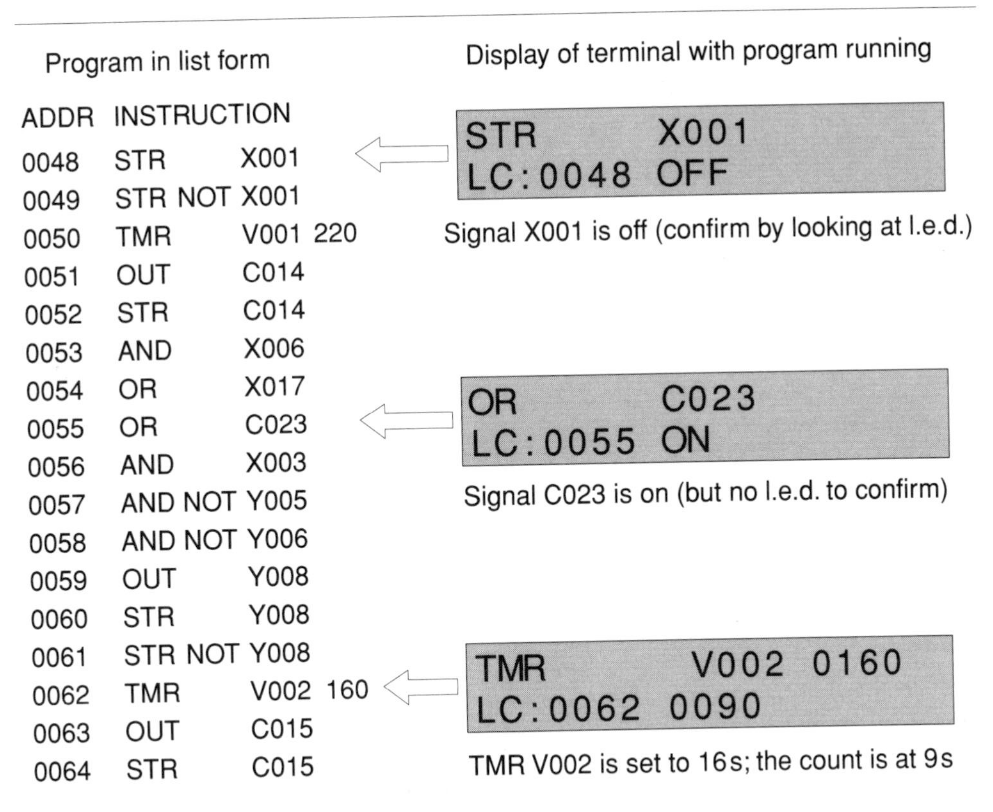

Figure 12.3. Visualizing a list program.

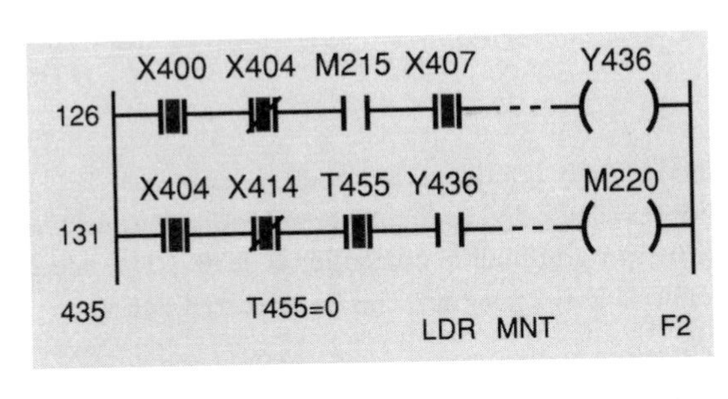

A ladder diagram as it appears on the screen of a Mitsubishi GP-80 handy graphic terminal, in monitor mode.

At the moment, signal M215 is absent and this is holding off output Y436.

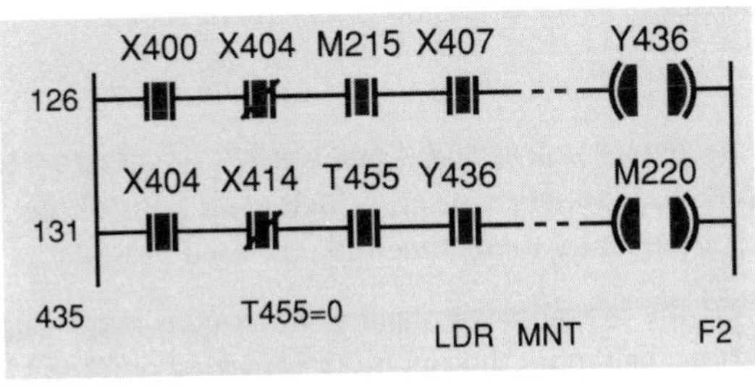

Now, signal M215 is present so output Y436 is on. As a result, the 'contact' belonging to Y436 in the next rung is closed, enabling M220 to be turned on.

Note that timer T455 is timed out; its 'contact' is closed and its remaining time is 0 s.

Figure 12.4. Visualizing a ladder program.

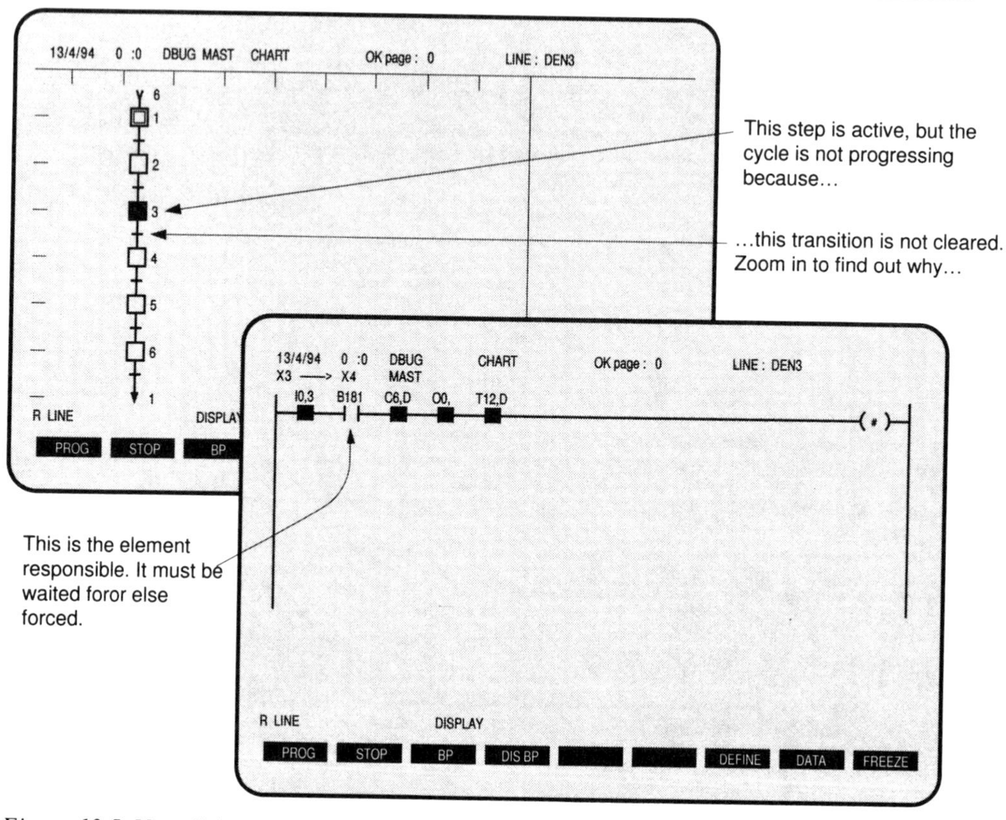

Figure 12.5. Visualizing a sequential function chart program.

12.4 Commissioning

When all pre-commissioning tests are satisfactorily complete, the plant is ready for commissioning. The programming terminal is connected to the programmable controller and set to visualize the program. The CPU of the programmable controller is set to RUN and the plant is operated on a trial basis. Modifications to the program can be expected due to

- inaccurate specification
- incorrect interpretation of the specification
- program error
- plant alterations.

Of course, wiring changes should not be necessary unless additional sensors or actuators are involved. The programming terminal can be used to write, insert or delete sections of the program as required (CPU stopped) and subsequently to document the revised program.

The purpose of commissioning is to 'prove' the overall design and to iron out as many bugs as possible in advance of full-scale operation. The more thoroughly it is carried out, the greater will be the rewards in terms of reliable operation and the avoidance of disruptions due to faults.

12.5 Fault-finding

Faults are inevitable in any plant, but in an automated plant the effect of a fault tends to have greater consequence; production may be completely stopped as a result of a very minor defect. Fault-finding is therefore an important skill in the automated plant, and one for which the programmable controller itself can be used as a tool.

At the outset the fault-finder should be reasonably equipped for the task; this means having the relevant documentation to hand (connection diagrams, I/O schedules, program printout, cross-reference, etc.). The source of the fault can best be identified by systematically eliminating the most likely culprits. Most new users, perhaps naturally, suspect the programmable controller first when any fault occurs. But Fig. 12.6 shows that this is unjustified: the weight of suspicion should fall on items *outside* the programmable controller.

It is normal fault-finding practice to check visually the items of plant associated with the problem before using any test instruments or tools. If this simple check yields no clues, then more serious work can be done. Table 12.1 lists the more common faults in the automation system, with their symptoms and possible remedies. It is by no means exhaustive, but it offers practical guidance for maintenance staff engaged in this activity. The table takes no account of the diagnostic functions of the controller itself. These functions can be extremely helpful to the fault-finder, especially in the case of faults *within* the controller. The reader is referred to the manufacturer's literature on this point.

Most true programmable controller faults are likely to be in the I/O modules and the power supply. Diagnostic functions or fault lamps on the controller may be able to confirm this kind of fault without further investigation. Most controllers recognize and report the following faults:

- CPU fault
- memory fault
- low battery
- I/O module fault (internal)
- watchdog timer overrun.

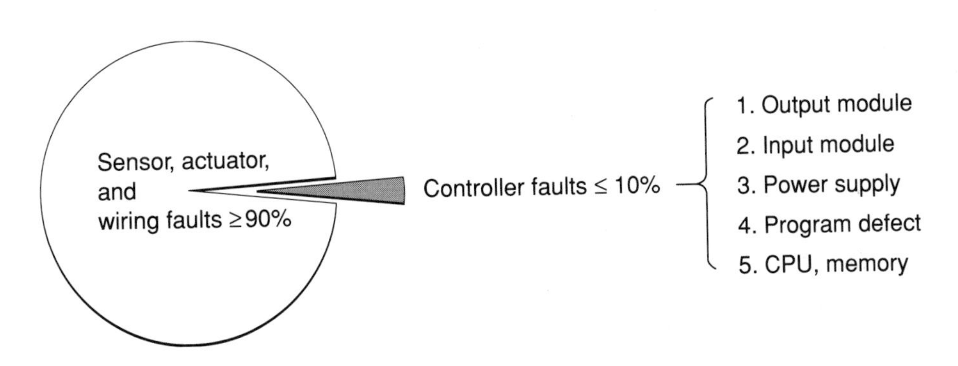

Figure 12.6. Approximate share-out of faults in an automated system.

Table 12.1. Fault-finding guide

Symptom	Likely cause	Remarks
Total failure or stoppage of the automation system	Power failure	Every area of the plant will be affected
	Supply turned off	If power lamp is off, check the main switch or circuit-breaker
	Emergency stop device activated	Inspect each emergency device and reset if it safe to do so
All inputs fail	Short circuit or earth fault on sensor or wiring	Consider stopping program. Disconnect inputs one by one until the input voltage returns
All outputs fail	Control circuit fuse or miniature circuit-breaker (mcb) operated	Test the control circuit supply using a voltmeter. Repair or reset as necessary
	Program stopped	Run lamp not on; CPU/MEM lamp may indicate an error. Secure safety functions and restart program
Group of outputs fail	Fuse or mcb for that group of outputs has operated	Output LEDs on: short circuit or earth fault may be the cause; check for voltage and repair or reset as necessary
	Faulty output module	Output LEDs are off: replace module using recommended procedure
Single output device fails to turn on	Wiring fault	Output LED is on, voltage at output is normal and voltage at device is absent
	Device fault	Output LED is on, voltage at output is normal and voltage at device is normal
	Absent input signal	Find out which one by reading ladder and checking LEDs or by monitoring the program

Table 12.1. Fault-finding guide (continued)

Symptom	Likely cause	Remarks
Single output device fails to turn on	Module fault	Confirm that all signals for turning on the output are correct. Check output LED and output voltage. Replace module, etc.
Output fails to turn off	Control signals are keeping it on	Find out which one by reading ladder and checking LEDs or by monitoring the program
	Module switching device destroyed	Output LED is off but the output itself remains on (relay contacts welded, transistor or triac faulty). Replace module
Program stopped	Mode set to STOP by hand or by program	Set to RUN if safe to do so
	CPU or MEM fault	Fault lamp confirms this. Reload good copy of program and set to run. Use diagnostic functions to check scan time; if max. value has been exceeded analyse program to find a means of reducing it
No program	Loss of power to RAM memory	Low battery lamp is on. Replace battery, reload program
Intermittent fault or spurious operation	Gross electromagnetic interference or lightning damage	Shield controller from possible sources of interference; check over-voltage protection and reload program

Good commissioning is meant to reveal and eliminate defects in the program. Occasionally, however, such defects come to light after many years of operation, and even then only due to some obscure and unforeseen circumstances. Finally, faults in the CPU or the memory are quite rare and can immediately be identified by the relevant indicator lamp on the controller.

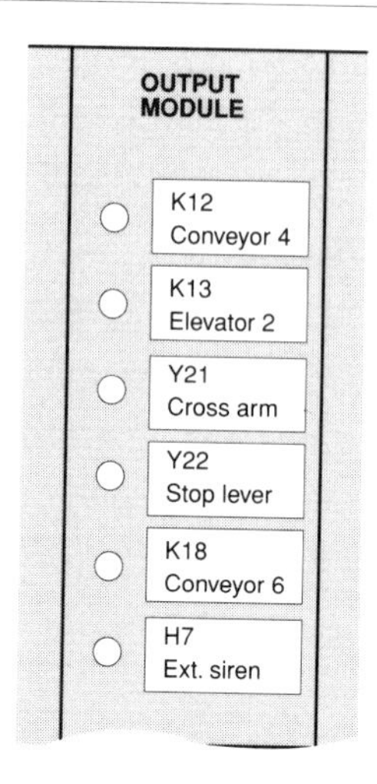

Figure 12.7. Marking the I/O name can be helpful for fault-finding.

Most manufacturers include a troubleshooting chart with the user manual, and the use of this chart is recommended for the positive identification of hardware faults.

The lot of the fault-finder can be improved by the following simple practical measures.

1. Each input and output terminal can be marked with the name of the connected device; this eliminates having to read the I/O schedule or connection diagram to identify the source of the signals (Fig. 12.7).
2. The field wiring terminal racks can be arranged in logical groups that are clearly identified (e.g. one rack for inputs, one for outputs, etc.) and readily accessible; each terminal should serve one purpose only, should carry a unique number and its function should be clearly recorded on the wiring diagrams. These arrangements ensure that the fault-finder has quick access to the correct terminals for test purposes.
3. The program can be designed to make use of *replies* to generate fault alarms. A reply is an input signal which verifies that an output has done its job properly. Figure 12.8 gives two examples of this.
   - Output Y7 controls a motor through a contactor (Fig. 12.8(*a*)). When the output is turned on, all the contacts of the contactor should close immediately; the motor should run and input X5 should get a signal. The program looks for this signal within 1 s of turning on Y7. If it does not arrive on time, alarm output Y15 is turned on. We describe this as an indirect reply because the input signal reports on the contactor rather than on the motor itself.
   - Output Q2.7 controls a cylinder through a control valve (Fig. 12.8(*b*)). When the output is turned on, the valve extends the cylinder. When the cylinder is fully extended it activates the proximity sensor, which in turn gives a signal at

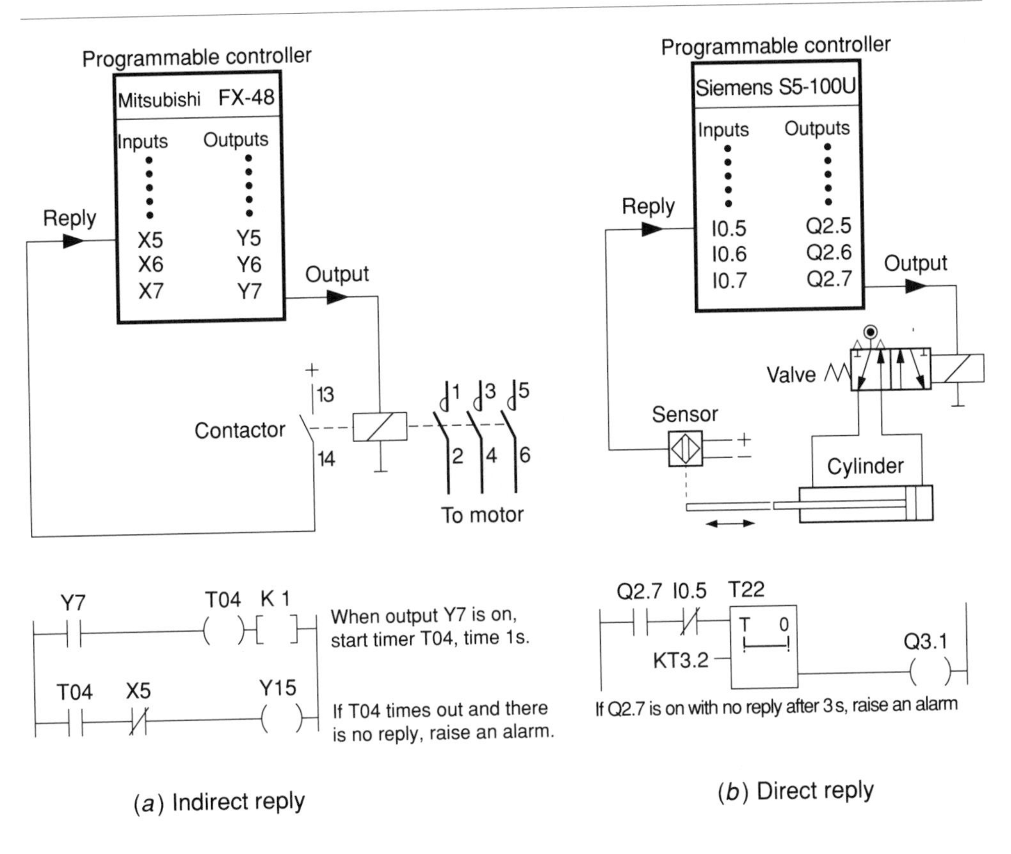

Figure 12.8. Obtaining and using replies to generate alarms.

input I0.5. The program looks for this signal within 3 s of turning on Q2.7. If it fails to appear, an alarm output Q3.1 is turned on.We call this a direct reply because the input signal is derived from the controlled actuator (not from an intermediate device).

4. A message display can be used to signal fault conditions. This eliminates any doubt about the cause of stoppages. A model equipped with a printer interface (Fig. 12.9) permits the printing of the alarm events as they occur, in real time.

12.6 Forcing

It is very useful to be able to intervene manually in the operation of automatic machinery or plant, especially during commissioning or fault-finding. Figure 12.10 shows some of the techniques used in traditional electrical or pneumatic control.

Figure 12.10(a): A wire link (or 'jumper') is used to short-circuit the terminals to which a switch or sensor is connected. This action may prove conclusively that the switch/sensor or associated wiring is faulty. It may also enable the resumption of production until the fault is rectified.

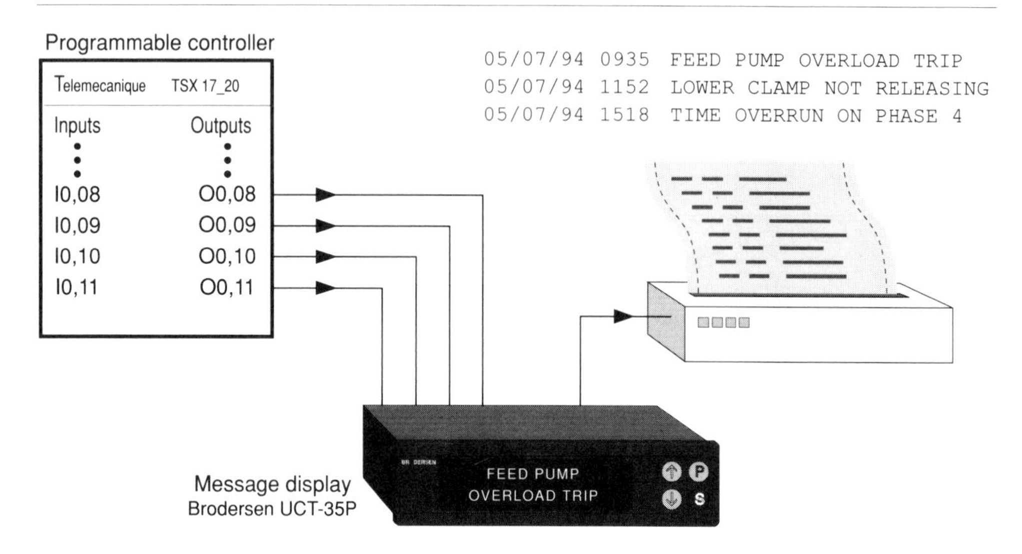

Figure 12.9. Generating a printed record of faults or alarms.

Figure 12.10(b): A contactor or relay is closed manually. This may be used to test the functioning of the coil circuit or the direction of rotation of a motor.

Figure 12.10(c): The override button on a control valve is pushed. This may be used to drive a cylinder to a particular position, perhaps to restart a cycle.

The equivalent action in a programmable controller is called *forcing*. Forcing allows us to unconditionally turn on or off any element accessible to the controller—inputs, outputs, timers, counters, function chart steps, etc. The only tool needed is the programming or adjustment terminal.

Figure 12.11 shows how forcing is carried out with the Sprecher+Schuh PRG-20 programming terminal. The user must decide whether to force the element on or off. In our example, output Y005 is forced on and internal relay C008 is forced off. The element's new status is indicated by 'fON' or 'fOFF' as appropriate. A forced element is released by 'unforcing' it—not by forcing the opposite status on it.

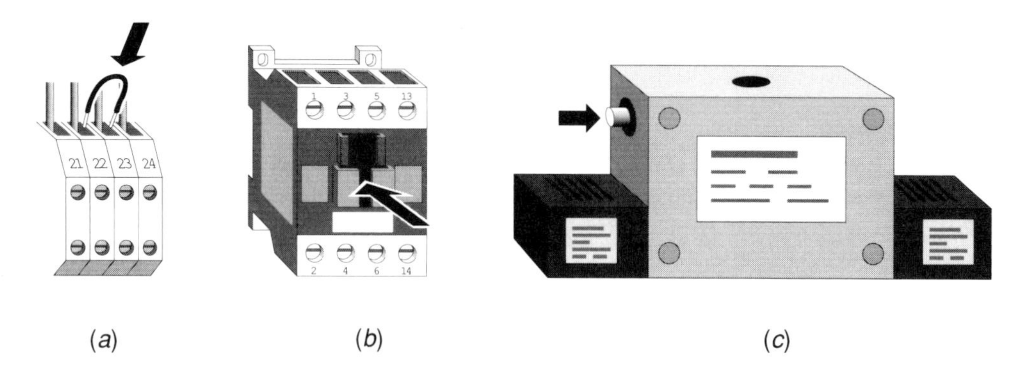

Figure 12.10. Examples of forcing in electrical or pneumatic control systems.

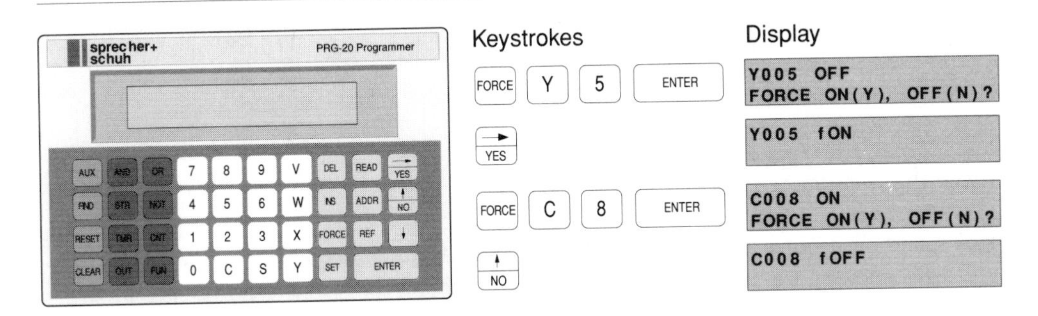

Figure 12.11. Forcing with a programmable controller.

The consequences of forcing can be significant, even dangerous, since hardware or software controls are being deliberately bypassed. Thus the user must

- plan the actions carefully and try to envisage all the possible consequences
- never compromise the action of safety devices
- follow the manufacturer's recommended procedures strictly
- record and display the changes made and have a schedule for unforcing
- avoid forcing in situations of extreme pressure or in case of doubt.

12.7 Exercises for Chapter 12 (Solutions in Appendix 1)

12.1. Why should pre-commissioning tests be carried out?

12.2. Explain the value of the monitoring facility during fault-finding.

12.3. In the car-park project (Chapter 10) the entry barrier fails to rise when expected. Explain how the source of this fault can be located. (Assume a graphic programming terminal is available.)

12.4. Describe the symptoms of a total failure of the input power supply for a programmable controller and the likely cause(s) of such a fault.

12.5. Explain how forcing can be used to confirm the failure of a remote sensor.

Appendix 1 Solutions to exercises

Chapter 1

1.1. Because it is the decision-making part.
1.2. The program.
1.3. To the input terminals.
1.4. Because they have few (if any) moving parts, are constructed with high-quality components and are tested with automatic machines that can detect even potentially defective components.

Chapter 2

2.1. Any of the following: limit switch, proximity sensor, photoelectric sensor.
2.2. 115 V, 230 V and 400 V a.c. for relay and contactor circuits and 24 V d.c. for programmable controller inputs.
2.3. Power supply module, CPU module, input module and output module.
2.4. Twenty I/O (twelve inputs and eight outputs).

Chapter 3

3.1. Transistor type.
3.2. This would be done, presumably, to limit the number of inputs used on the programmable controller, but it has disadvantages:

- a faulty stop switch cannot be identified by normal monitoring
- logic changes may also mean wiring changes.

Thus in the interests of easier troubleshooting and flexibility it is preferable for each switch to be allocated its own input.

3.3. Two inputs (if you find this answer puzzling, refer to Fig. 3.2, p. 21).
3.4. To control a load whose voltage or current exceeds the rated values of the programmable controller's output.
3.5. Advantages are as follows.

- Messages are in clear text, and are easily surveyed in one location.
- Programmable for operator, maintenance or management information.
- Reduction in the number of inputs and outputs required for dialogue.
- A high level of dialogue with the operator is possible, sophisticated operations become more manageable.
- A printed record of events is possible with some models.

Chapter 4

4.1. The output is turned off.
4.2. (This requires an extra AND instruction immediately before the output.)

4.3. (The INVERSE AND function replaces the AND function.)
4.4. (The existing program remains but a new section is added: a signal from K1 and S4 turns on K2.)
4.5. Advantage: accepts changes immediately. Disadvantage: loss of power = loss of program.
4.6. Because it warns of the impending loss of program if the mains should fail.
4.7. When it is untouched.
4.8. No, provided the program 'fits' in the memory.
4.9. The turn-on instruction must contain no conflicts, and the best way to ensure this is to issue the command only once.

Chapter 5

5.1. No. We could have used just one and interpreted its signal as follows. Signal present = summer selected; no signal present = winter selected. Note, however, that no OFF position is available now.
5.2. Each pump circuit would be fitted with a thermal overload relay or circuit-breaker and the program would require modification to include their effects.
5.3. Each occurrence of an input signal from a thermostat would have to be inverted, i.e. the opposite to those suggested in our solution.
5.4. (Authors' note: Only three of the controllers available to us were fitted with the real-time clock option and their programming procedures differed substantially. We are therefore leaving it to the reader to consult the manufacturer's manual to solve this exercise.)

Chapter 6

6.1. A retentive relay has its signal stored in a portion of the memory backed by the battery. On power return the retentive relay immediately assumes the state it held before the supply failure. It can be used to prevent disruption of the process due to a power failure.
6.2. Latching means 'locking'. The flip-flop always keeps its current state until it is deliberately reversed; thus it is SET until it is RESET and vice versa.
6.3. (This requires the use of two timers, say T1 and T2. T1 runs via the inverse of T2. T2 runs via T1. T1 is programmed with 2 s and T2 with 1 s.)
6.4. A sequential function chart gives a clear structure to the control sequence. A function chart is a standardized form of representation (IEC 848/1131-3). Accessible to several disciplines and levels of staff.
6.5. The solution is shown in Fig. Ex 6.5.
6.6. Figures Ex 6.6(*a*), (*b*) and (*c*) show the UK traffic lights' sequence.

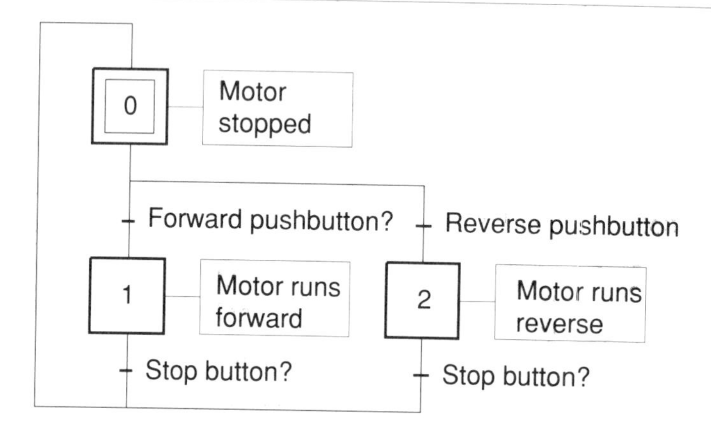

Figure Ex 6.5. Sequential function chart for reversing control of a motor.

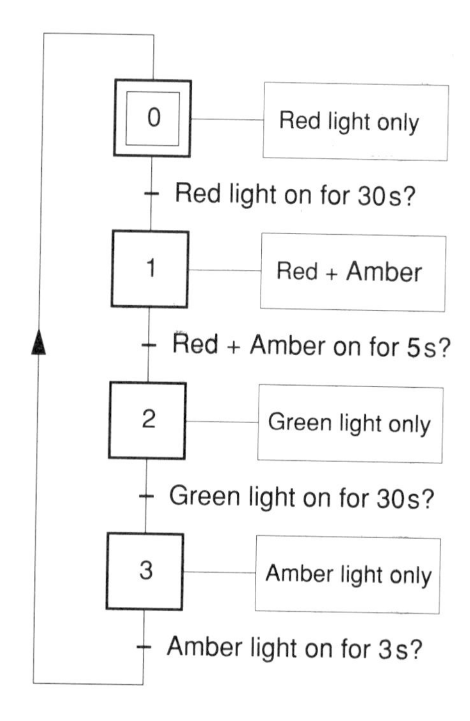

Figure Ex 6.6. (a) Sequential function chart for UK traffic light sequence.

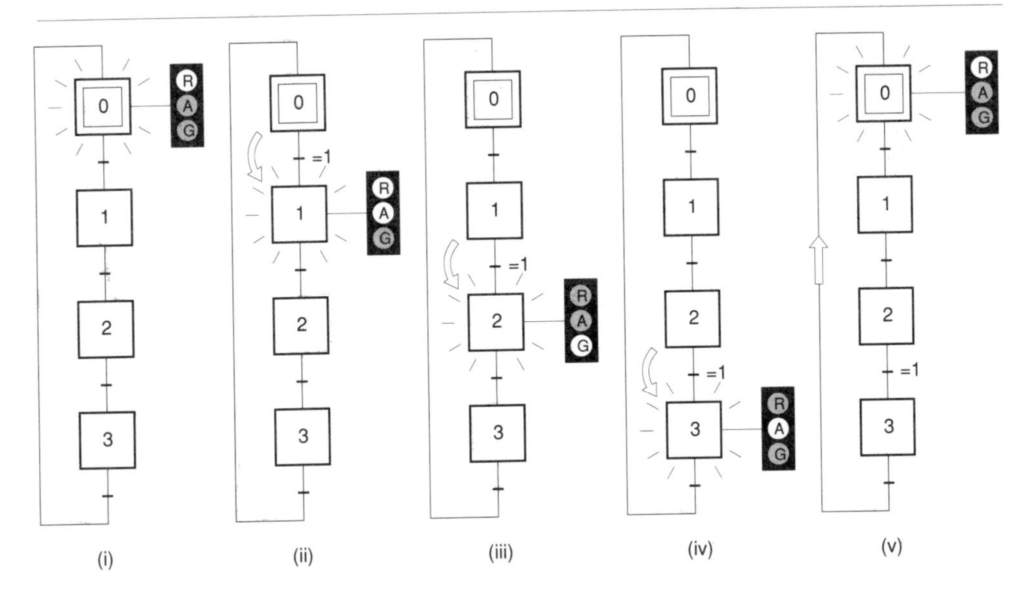

Figure Ex 6.6. (b) Visualizing the cycle for the UK traffic light sequence.

Chapter 7

7.1. (Use the general solution in Fig. 7.5 (p. 85) to help devise your program.)

7.2. (a) Introduce the new 24 V supply. (b) Assign the hooter to a different output address — one that is not already served by the 115 V supply. (c) Modify the program by replacing the old output address with the new one.

7.3. (Refer to Fig. 5.3 (p. 55) for details of the panel for an earlier project.)

7.4. Sonar or capacitive proximity sensors could be used. These are mostly three-wire types, and this means an additional negative conductor is provided for each one.

Chapter 8

8.1. The solution requires a modification of the control logic for the door lamps to enable them to flash. One method is shown in Fig. Ex 8.1.

MEM1 is turned on by a call or dispatch button if an access door is open; it then holds itself on until the door is closed again.

MEM1 is used to turn on MEM2 through a flasher (which may be ready-made within the controller or be assembled using two timers). When MEM1 is on, MEM2 flashes.

MEM2 is now placed in series (ANDed) with each of the lamp outputs to ensure that they flash on and off as specified.

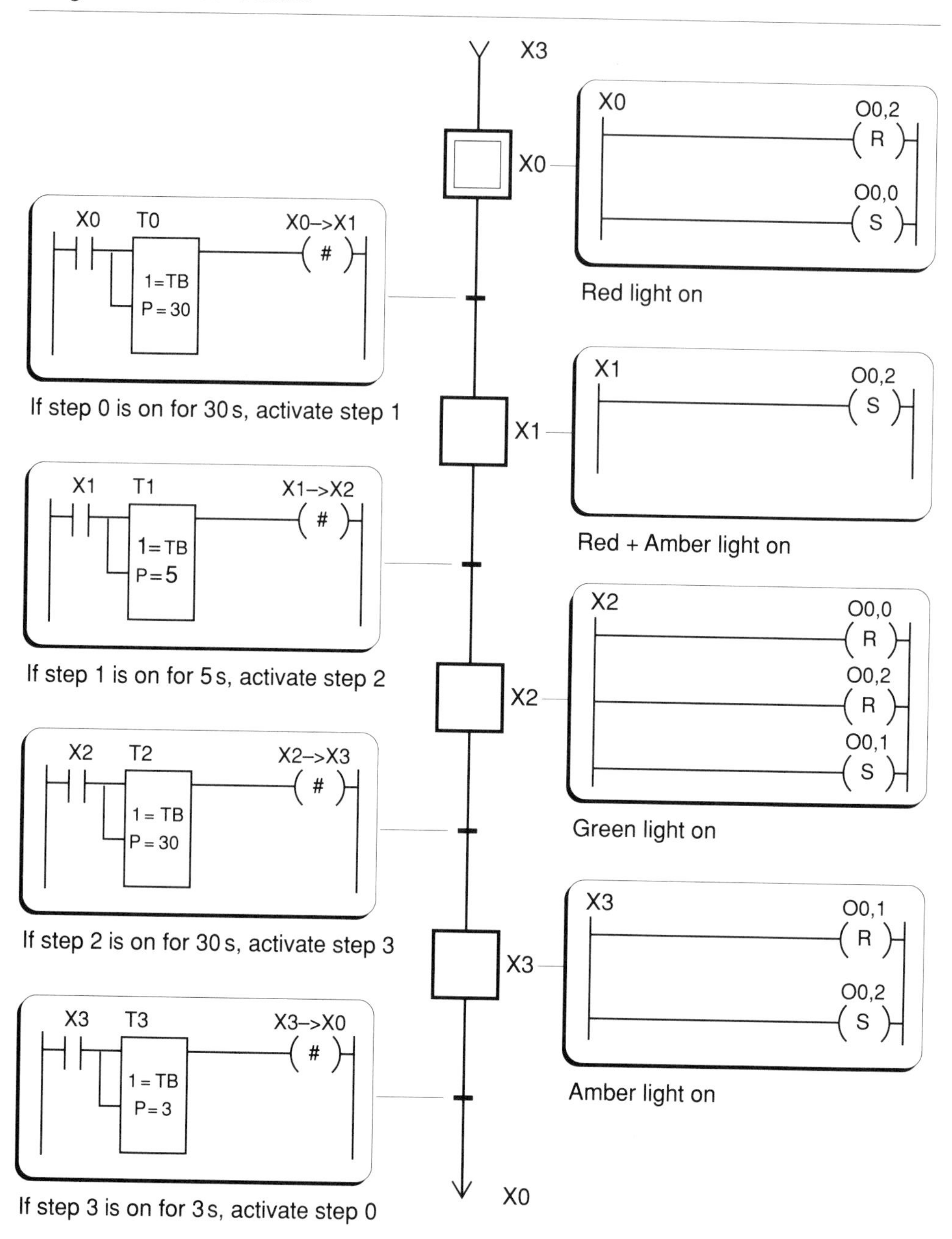

Figure Ex 6.6. (c) Solution for Telemecanique controller.

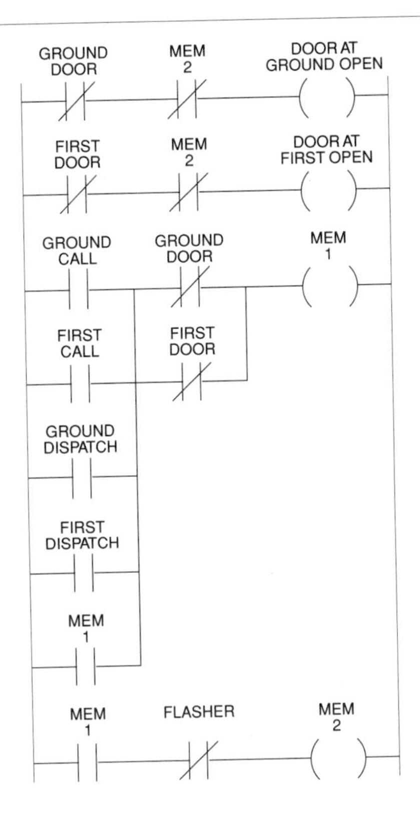

Figure Ex 8.1.

8.2. The program for the other controllers will be very similar to those given. The main differences will be the I/O addresses (refer to Tables 8.1 and 8.2, p. 94). Mitsubishi timers are: T50...T57, T450...T457, T550...T557 and T650...T657.
Siemens timers are: T0..T127 (CPU 103). Refer to Section 6.4 for examples.

8.3. The lamps must be connected to outputs that can be served with a 24 V common supply. Possibilities are as follows:

- Sprecher and Schuh: Move lamps to Y005...Y008. Change C1 to 24 V common. Move bell to Y003.
- Telemecanique: Lamp outputs unchanged; C3 and C4 changed to 24 V common. Bell transferred to O0,0.

Program must be rewritten in every case to account for the I/O alterations.

8.4. The lift stops and the brake is applied. Recovery is not automatic.

Chapter 9

9.1. Mechanical damage: use a suitable enclosure.
Ingress of dust or moisture: use a suitable enclosure.
High ambient temperature: choose a cool site or install cooling equipment.

9.2. Voltage rating of the cables must be ≥ 230 V a.c.
9.3. A bus is a common set of conductors used for transferring the I/O signals within the programmable controller using unique addresses. A small bus can replace a vast number of interconnecting cables.
9.4. Safety devices must be 'fail safe' and must not depend on the programmable controller for effective operation. The contact performing the safety function must be connected in the output circuit.

Chapter 10

10.1. Refer to Tables 10.1 and 10.2 (page 115) for the corresponding addresses. The ladder diagram will be broadly similar to those already given. The counters will be somewhat different (refer to Section 6.5 for details). Mitsubishi counters: C60..C67, C460..C467, C560..C567 and C660..C667. Sprecher +Schuh counters: CNT V001...V064.
10.2. A pressure switch must be fitted to the air system and its contact wired as an input. There are many possibilities for using this input within the program e.g. as a master control device (partial or total shutdown) or to raise an alarm.
10.3. A clock switch must be fitted and wired as an input. Possible extra logic: the entry barrier can be raised only if the clock input is present; when the clock input is absent, the exit barrier remains up until the car-park is empty.
10.4. A photoelectric sensor to be fitted beneath the barrier and wired as an input. The barrier's descent is not permitted until this sensor gives a 'clear' signal.
10.5. Write out your solution and compare it with Table 5.3 (page 64).

Chapter 12

12.1. To prove the correct functioning of sensors, actuators and the wiring.
12.2. Monitoring enables the fault-finder to see the state of all the signals relevant to the turning ON of outputs, internal relays, etc.
12.3. If the output for Y1 is OFF, use the monitoring facility to find the cause. If it is ON, check the voltage at Y1. Voltage present: defective valve, defective cylinder or no air. No voltage: wiring fault or control circuit supply fault.
12.4. A malfunction coinciding with all input signals absent. Caused by earth fault or short circuit in the input wiring, a blown fuse or a defective power supply.
12.5. Assuming the sensor fails to give a signal when it should, check first the relevant l.e.d. on the controller and then the status of the input signal with the programming terminal. If both of these confirm the absence of the signal, force ON the input. If, as a result, there is proper functioning, the sensor or the associated wiring has failed. Assuming the sensor fails to switch off when it should, check first the relevant l.e.d. on the controller and then the status of the input signal with the programming terminal. If both of these confirm the presence of the signal, force OFF the input. If, as a result, there is proper functioning, the sensor has failed or the associated wiring has a short circuit.

Appendix 2 Further reading

Berger, H., *Automating with the Simatic® S5*, Siemens Berlin, Munchen, 1989.

Bouteille, D. *et al.*, *Programmable automation systems*, CITEF Rueil-Malmaison, 1990.

IEC 1131-3, *Programmable controllers Part 3: Programming languages*, IEC, Geneva, 1993.

Johannesson, G., *Programmable control systems*, Chartwell-Bratt, Lund, 1982.

Moore, C. A. *et al.*, *Automation in the food industry*, Blackie, Glasgow, 1992.

Warnock, I., *Programmable controllers*, Prentice-Hall, Cambridge, 1988.

Appendix 3 Summary of changes driven by IEC 1131-3

Location identifiers

Prefix	Meaning
I	Input
Q	Output
M	Memory

Size identifiers

Prefix	Meaning
B	Byte (8 bits)
W	Word (16 bits)
D	Double word (32 bits)

The following examples show how these identifiers are used. The prefix '%' indicates that the elements are directly addressed.

%I48	Input bit 48
%Q19	Output bit 19
%IW7	Input word 7
%QB31	Output byte 31
%M255	Memory bit 255
%MW66	Memory word 66

Ladder contacts

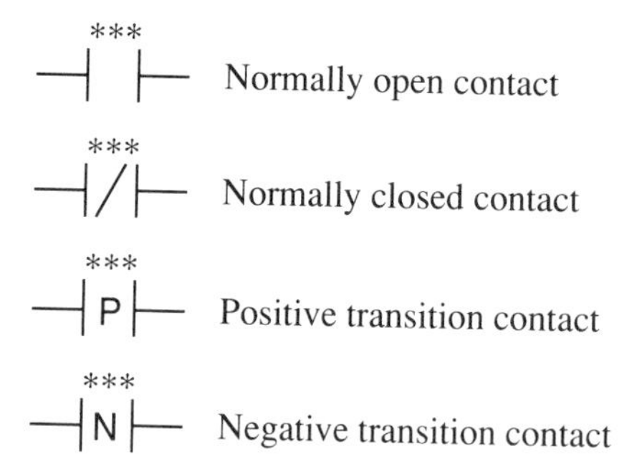

Ladder coils

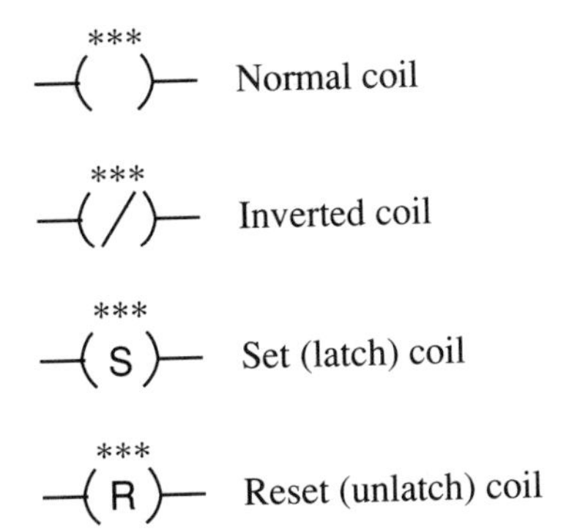

Figure A3.1 shows a ladder example.

Figure A3.1. Ladder example.

Table A3.1 shows the relationship between ladder and list languages.

Table A3.1. The relationship between ladder and list languages.

Ladder symbol	IEC 1131-3 mnemonic	Siemens S5-100U	Mitsubishi F_2-40/Fx	Sprecher+Schuh SESTEP® 290	Telemecanique TSX 17_20
*** |—| |—	LD	A	LD	STR	L
*** |—|/|—	LDN	AN	LDI	STR NOT	LN
*** —| |—	AND	A	AND	AND	A
*** —|/|—	ANDN	AN	ANI	AND NOT	AN
*** └| |┘	OR	O.	OR	OR	O
*** └|/|┘	ORN	ON	ORI	OR NOT	ON
*** —( )—	ST	=	OUT	OUT	=
*** —(/)—	STN				=N
*** —(S)—	S	S	S		S
*** —(R)—	R	R	R		R

Index

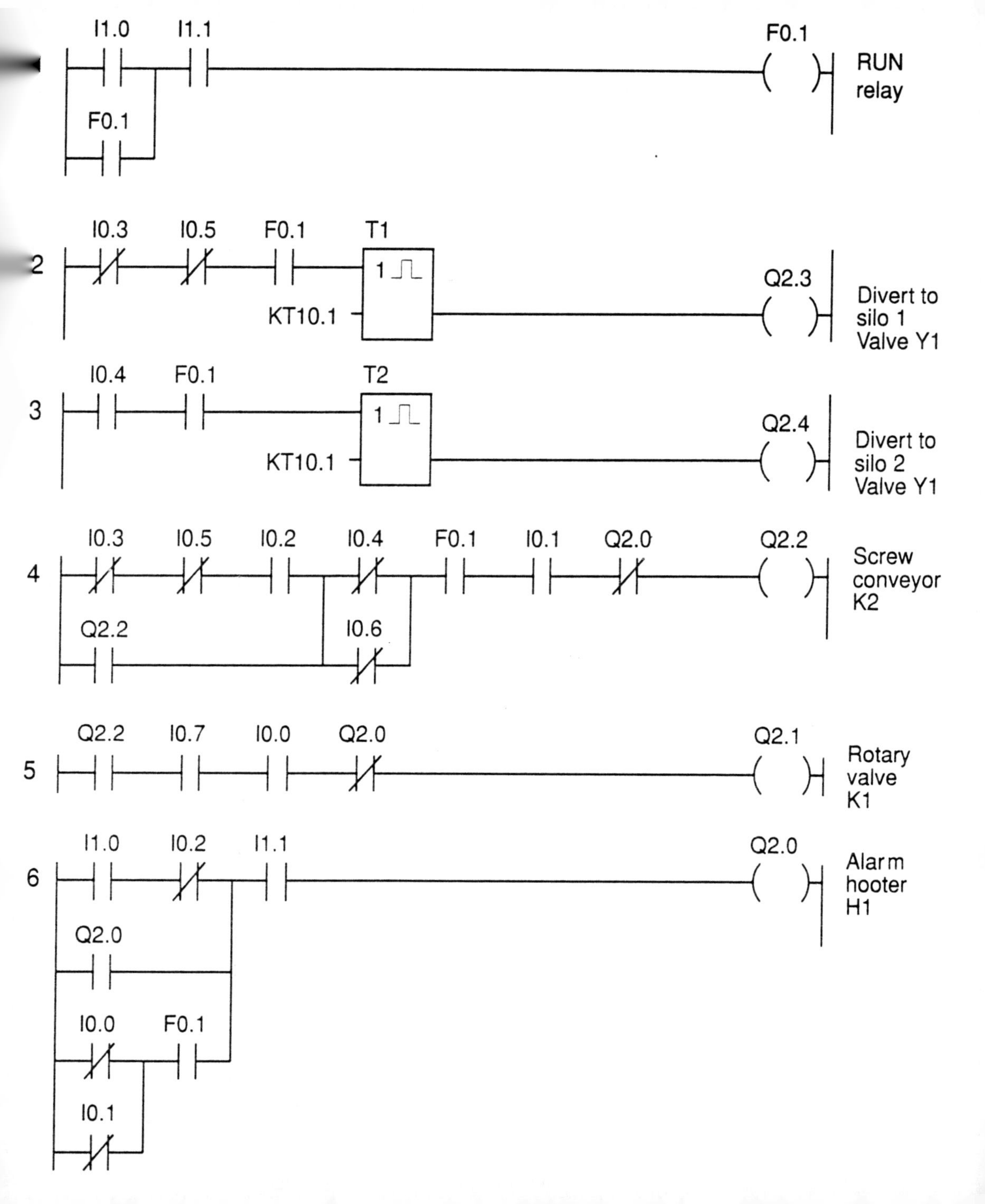

I1.0
I1.1
F0.1
RUN relay
F0.1
2
I0.3
I0.5
F0.1
T1
1
KT10.1
Q2.3
Divert to silo 1 Valve Y1
I0.4
F0.1
T2
3
1
KT10.1
Q2.4
Divert to silo 2 Valve Y1
I0.3
I0.5
I0.2
I0.4
F0.1
I0.1
Q2.0
Q2.2
4
Screw conveyor K2
Q2.2
I0.6
Q2.2
I0.7
I0.0
Q2.0
Q2.1
5
Rotary valve K1
I1.0
I0.2
I1.1
Q2.0
6
Alarm hooter H1
Q2.0
I0.0
F0.1
I0.1